S0-AFS-399

EXTREME FEAR

EXTREME FEAR

THE SCIENCE OF YOUR MIND IN DANGER

Jeff Wise

palgrave
macmillan

EXTREME FEAR
Copyright © Jeff Wise, 2009.

All rights reserved.

First published in hardcover in 2009 by
PALGRAVE MACMILLAN®
in the US—a division of St. Martin's Press LLC,
175 Fifth Avenue, New York, NY 10010.

Where this book is distributed in the UK, Europe and the rest of the world,
this is by Palgrave Macmillan, a division of Macmillan Publishers Limited,
registered in England, company number 785998, of Houndmills,
Basingstoke, Hampshire RG21 6XS.

Palgrave Macmillan is the global academic imprint of the above companies
and has companies and representatives throughout the world.

Palgrave® and Macmillan® are registered trademarks in the United States,
the United Kingdom, Europe and other countries.

ISBN: 978–0–230–10348–1

Library of Congress Cataloging-in-Publication Data

Wise, Jeff.
    Extreme fear : the science of your mind in danger / Jeff Wise.
       p. cm.
    Includes bibliographical references and index.
    ISBN 978–0–230–61439–0
    (paperback ISBN 978–0–230–10348–1)
       1. Fear. 2. Neuropsychology. I. Title.

BF575.F2W565 2009
152.4'6—dc22   2009018929

A catalogue record of the book is available from the British Library.

Design by Newgen Imaging Systems (P) Ltd., Chennai, India.

First PALGRAVE MACMILLAN paperback edition: February 2011

10 9 8 7 6 5 4 3 2

Printed in the United States of America.

To Sandra, and to Rem, our joy and terror

CONTENTS

ACKNOWLEDGMENTS

I OWE A DEBT of gratitude to my agent, David Kuhn, whose persistence was essential in bringing this project to fruition; and also to my editor, Luba Ostashevsky, who, with skill and enthusiasm, helped me take the manuscript to a level beyond what I could have achieved on my own.

I would also like to thank the magazine editors who worked with me on fear-related articles as I honed my understanding of the topic: Kent Black at *Outside's GO*, Chris Raymond at *Details*, and David Dunbar at *Popular Mechanics*.

Finally, I could never have written a word of this book without the support of my wife, Sandra Garcia, who, in addition to emotional and intellectual support, kept our newborn baby pacified while I sweated under deadline.

INTRODUCTION

THE MYSTERY OF FEAR

ON JUNE 3, 1970, shortly before noon, a British pilot named Neil Williams strapped himself into the harness of his blue-and-white Zlin Akrobat, a rugged but nimble single-engine airplane built in Czechoslovakia. The World Aerobatic Championship was coming up, and Williams planned to prepare himself by running through the sequence of maneuvers that he'd be flying in competition.

With a lantern jaw, deep-set eyes, and shock of dark hair swept back from a high forehead, Williams looked every bit a casting director's idea of a daredevil pilot—and in his case, looks did not deceive. Williams was a veteran flyer with a vast and varied store of experience under his belt. In the course of his career he had flown more than 150 different kinds of airplanes and accumulated more than six thousand hours in flight time. A retired Royal Air Force test pilot, and four-time winner of the UK aerobatic championships, he was, at thirty-six, already regarded as one of the greatest all-around pilots that Britain had ever produced. But his skills had never been tested as they were about to be.

Rafts of fair-weather clouds drifted over the Royal Air Force base at Hullavington, England, as Williams lined up on the runway, opened the throttle, and roared into the air at full power. The wind aloft was gentle, and as Williams climbed he noted with satisfaction that there was no detectable turbulence—that meant he'd be able to carve his maneuvers all the more precisely.

Williams ran his sequence twice through without incident, then brought the Zlin back to level flight and prepared to practice his routine one final time. After only a few minutes in the air, he was already near his limit for fatigue. Competitive aerobatics is a uniquely demanding undertaking. As a mental discipline, it requires exacting attention to detail, the ability to think quickly and three-dimensionally, and the ability to maintain one's poise while rapidly moving through maneuvers that turn the plane upside down, cause it to fall backward, or spin like a top. As a physical discipline, it requires grit and superb fitness as the airplane's abrupt changes of direction slam the pilot from one side of the cockpit to the other, with centrifugal forces at times pressing on his body with nine times its actual weight and at other times leaving him hanging upside-down from his harness straps. A four-minute aerobatic routine is enough to leave a pilot drained and soaked in sweat.

Midway through the third run-through of his routine, Williams was coming over the top of a loop, a high arcing figure in which his plane carved through the air like a high fly ball. As it reached the top, Williams was upside-down in his seat, the checkered farmland of southwest England arrayed above his head, the cloud-dappled sky under his seat. The plane continued its arc downward past its apex, and the horizon sank toward the bottom of his windscreen until all he could see in front of him was ground. His descent grew steeper and steeper until he was staring straight down.

Barreling earthward through fifteen hundred feet, Williams hauled the stick toward his chest in order to pull the Zlin back to level flight. He clenched his abdominal muscles in anticipation of the resulting g-forces, as gravity combined with the centrifugal force of the plane's curving path would press him into his seat with five times his normal weight. Only by grunting and clenching his leg and stomach muscles could he prevent the blood from rushing out of his head and causing him to black out.

The plane was just coming level with the ground, one thousand feet up, when—

BANG!

A jolt shook the airplane. The Zlin started rolling left—all except the left wing, which stayed oddly level with the horizon. Williams instantly intuited what had happened: The force of the pull-out had likely broken the internal spar that gave the wing most of its strength. If that were the case, then the whole wing was about to fall off. He pushed the stick all the way to the right, but the plane kept rolling left. The ground was just 300 feet below and rising fast.

For most pilots, that would have been the end. But in the few seconds he had left before his plane cratered, Williams had an insight. He remembered the story of a Bulgarian pilot who had suffered a malfunction in a similar Zlin model years before. The circumstances in that case had been different—the Bulgarian had been flying inverted when a bolt failure in one of the wings had caused the plane to unexpectedly flip right-side up. But a detail of the story stuck out: Once the Bulgarian's plane was right-side up, the wing had snapped back into place. Maybe Williams' situation was analogous, but reversed. If he went from right-side up to upside down, his wing might snap back in place, too.

In less time than it takes to form a complete thought, Williams threw the stick hard to the left until the Zlin was fully inverted, then pushed the stick forward. His face swelled and turned red as gravity and centrifugal force drained blood from his body into his head.

WHUMP!

With a satisfying thud the wing settled backed into place. By now Williams was almost in the treetops, and for a moment he was sure he was going to crash. Then the plane began to climb.

Hanging in his harness, Williams coaxed the stricken craft skyward, eking out precious altitude foot by foot. He didn't have much time: His engine, he knew, would only run for eight minutes upsidedown. Without a parachute, his options were stark. Should he try to crash land upside down in the trees? Find a lake to ditch in?

Just then the engine sputtered and died—a new potentially fatal disaster. Williams scanned the cockpit and quickly found the problem.

In the initial jolt, he had accidentally hit the knob that shuts off the fuel supply to the engine. He flipped it back to the "on" position. After a few coughs, the engine came back to life.

Williams was running short on time. He decided his best chance for survival was to crash-land at the airfield. He guided the Zlin home and set up his landing approach upside down. As the end of the runway passed above his head, he pushed the stick hard to the right and rolled the plane right-side up. Again the left wing folded up and the plane careened sideways as it touched down. Williams curled into a ball until the plane stopped moving, then broke open the damaged canopy and leapt free. The plane was a wreck, but he had survived without a scratch.

HOW WILLIAMS MANAGED to survive the catastrophic failure of his wing at low altitude is a mystery. In aeronautical terms, to be sure, the question of staying in the air was simply a matter of physics. But the psychology of what happened is another matter. By conventional understanding, Williams should have died that day. Under such intense pressure, with fatal impact a few seconds away, the surge of hormones should have been so intense, the neurons of his fear circuitry so overloaded, that Williams should have been barely able to function, let alone come up with a creative solution in the blink of an eye.

Something extraordinary must have been going on in his brain. Some mechanism in his psychological tool kit must have somehow protected him from panic. Perhaps it even gave him an extra dose of mental power to get him through the crisis. Whatever he possessed, it was a rare talent. Rare, but not unique. The annals of human achievement are peppered with stories of people who managed to survive lethal danger by thinking on their feet. How do they do it? What makes them different? And, most importantly, what can the rest of us learn from them?

I first became interested in the subject of extreme fear when the focus of my career as a magazine journalist gradually shifted toward adventure travel. One adrenaline-cranking assignment led to another, increasing each time in danger and intensity. A heli-fishing article in

British Columbia led to a fly-in fishing camp in Alaska, which led to an Alaskan bush-flying story, which led, eventually, to my getting my pilot's license. Then I got my glider license, which led to a story about flying powered hang-gliders low over the New Mexico desert, which led to another that involved aerobatics in a World War II fighter. Along the way, I swam down whitewater rivers, slept in snow caves, rode in homemade submarines, and rappelled down cliffs. The more I demonstrated my willingness to take on extreme adventure, the more editors called on me for that sort of assignment.

As my work life took me ever more frequently into hair-raising situations, I found myself ever more often at the limits of what I could handle. And then one day an assignment brought me to The Bungy Zone, a bungee operation near Nanaimo, British Columbia. The jump-off point was a bridge 150 feet above a rocky, deeply shadowed gorge. As I watched, an enthusiastic daredevil spread his arms and made a swan dive into the abyss. Next it was my turn. Even though I knew rationally that I wouldn't be hurt physically, I felt sure that I couldn't leap into that void without something in my mind snapping. But I had a job to do, so I stepped up onto the bridge. The feeling of dread grew ever more intense as I walked to the jumping platform, and then as I sat while the operator strapped the bungee cord around my ankles. The looming loss of control was palpable and terrifying— more terrifying, at that moment, than the physical act of jumping. I was more afraid, I realized, of what was going to happen in my mind than what was going to happen to my body. I had the overwhelming feeling that some strange, intense kind of madness was struggling to take over my mind. What, I wondered, inside my brain was creating this sensation?

When I got home I began researching the science of the fear response and found that the field was in the midst of a golden age. New tools were coming on line that revealed the workings of the brain in unprecedented detail. One of the most important was functional magnetic resonance imaging, or fMRI, a technique that allows researchers to directly visualize mental activity as it takes place deep within the brain.

As it turned out, of all the human emotions, fear is the one that has been studied in the greatest detail. As emotions go, it's perfectly suited for investigation. Conceptually, it's uncomplicated: a system that detects danger and responds so as to maximize an organism's probability of survival. Ancient in origins and ubiquitous in daily experience, fear can easily be generated in a laboratory setting in both human and animal subjects.

Fear can manifest in many ways, but all rely on the same underlying neurological system, a collection of processing centers that range from simple structures that first appeared hundreds of millions of years ago to sophisticated regions of the brain that evolved quite recently. These centers work together to coordinate our response to danger.

Because many of the brain regions work deep beneath the surface of consciousness, their workings remain hidden, mysterious, and often surprising. Fear can strike without warning, seeming to pop out of nowhere to hijack the mind. And attempting to control it can lead to paradoxical effects: the harder you try to suppress fear, the worse it can get.

Understanding how fear is generated in the brain is important not just for those who regularly risk their lives, but for all of us. We all have to deal with threats of one sort or another in our lives, whether it's a meeting with an angry client, finding yourself at the top of a triple-black-diamond ski run, or asking your girlfriend to marry you. No matter how you live your life, there are going to be times when your heart is pounding, your mouth is dry, and your hands are shaking.

In an ideal world, our nervous system would respond with perfect efficiency every time we're under pressure, jumping instantly into high gear whenever danger looms and lying dormant the rest of the time. But it doesn't work like that. Often, the response initiated by our fear centers is just as dangerous as the threat that spurred it in the first place, if not more so.

People in the grip of true terror can feel utterly hijacked. Soldiers throw down their guns and run away. Pilots lose control and crash their planes. In such cases the grip of fear feels like possession by some implacable alien force. Indeed, the word "panic" comes from the Greek god Pan, whom the classical Greeks believed could overtake travelers in lonely spots and send them suddenly running in blind terror. To the ancient mind, possession by a malign deity seemed the only plausible explanation for such behavior.

Few of us, hopefully, will have to endure such extreme fear. But even mild stress can affect our performance, as anyone who has ever stammered and sweated through a job interview can attest. When we're gripped by fear, we generally have two problems to deal with: the thing that we're worried about, and the fear that goes along with it.

Or to look at it another way: If we can learn to deal with our fear, we've instantly cut in half the number of things we have to worry about.

The good news is that we can do just that. By better understanding how our fear response works, we can take more effective steps to counter it. And that's exactly what *Extreme Fear* is about: helping you better manage your response to pressure by better understanding the underlying whys and hows.

In the course of this book, we'll first talk about how the mind and body respond to being in fearful situations, in ways that are both helpful and unhelpful. Next, we'll look in greater detail at three major categories of high-stress situations. Then we'll survey the major strategies available to us for coping with dangerous, fearful, and high-pressure situations. And finally, we'll round out the book's journey by coming to a more sophisticated understanding of the role fear plays in our lives.

Fear isn't all bad. Just as it can be the most intensely awful experience, it can also leave us feeling more exhilaratingly alive than ever before. By taking steps to prepare ourselves, we can do more than just hope for the best. Combining the latest research findings with time-proven techniques, we can train ourselves not only to hold up under pressure, but to excel.

PART ONE

FEAR IS A PARALLEL MIND

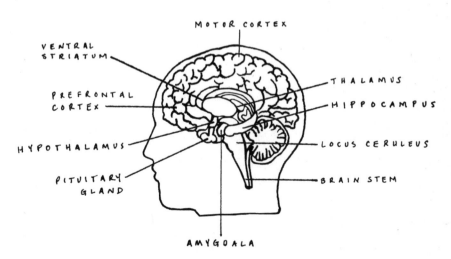

Some of the key centers involved in the fear response. The battle for self-control plays out in the dynamic between the prefrontal cortex and the amygdala.

Art by Sandra Garcia

CHAPTER ONE

THE PERSON THAT
FEAR MAKES YOU

THE PLANE IS at twelve thousand feet when the door slides open, and all at once the cabin is filled with a chill and the sound of roaring wind. Outside, the clear blue light of an autumn afternoon bathes the North and South Forks of Long Island. Without pausing, a man in a red jumpsuit crawls to the edge of the doorway and tumbles out, followed by a woman in yellow, then another man in blue. The tail of the plane bumps upward each time a jumper's weight leaves the sill of the door.

Now my instructor and I are alone here in the bare, seatless cabin. "Okay," he says. "Let's scoot forward."

I think: *Ugh.*

He starts to scramble forward, crab-like, and I have no choice but to follow suit, because we're strapped tightly together, his chest to my back, his pelvis against mine. I feel numb. My chest feels tight, my mouth is dry. I wish I were anyplace but here. My mind is racing as I try to keep control. I think of that hoary cliché of skydiving, the petrified first-timer clinging white-knuckled to the edge of the door. Am I going to be like that? Am I going to lose it? Or am I going to keep it together?

I feel weak, barely able to move my own weight. As we reach the door, a whole uninterrupted panorama of Long Island opens up before me, the forests dappled with the oranges and reds of early autumn. To the south, the arc of the Atlantic Ocean merges into the paler blue of the sky. Everything looks so crisp, so crystalline, with a beauty that would be stunning if I had the least interest in looking at it. The wind howls past as we put our feet over the edge. There's nothing below us but air, thousands and thousands of feet of empty air.

I think: *What the hell am I doing here?*

Actually, the answer is very simple. I've volunteered to take part in an experiment in the nature of fear. It's a decision that, at the moment, I'm regretting deeply.

Duncan makes one final check of the equipment. The electronic recorder strapped to my waist is on and collecting data from the sensor vest around my torso. Duncan's voice comes from behind me, the words that he'd explained on the ground would be our final signal to push away from the airplane: "Head back—ready—set—arch!"

Oh God.

GOING UP IN a skydiving plane for the first time taught me the lesson all over again: In the grip of fear, our minds work differently. Our bodies and brains simply don't respond the way we expect them to. Tasks that are very simple when undertaken in a state of calm become intensely challenging, or even impossible, when the adrenaline is pumping. Conversely, we might rise to the occasion and perform much better than we would have any reason to suspect. It's impossible to know beforehand which way we'll break.

The question of who will rally in the face of danger, and who will crumble, is an imponderable that has fascinated people since time immemorial. Some who talk tough turn into cowards in the face of battle; others who seem meek prove fearless. As the eighteenth-century French aristocrat François de La Rochefoucauld put it: "One cannot answer for his courage when he has never been in danger."

Twentieth-century psychologists, eager to figure out how to extract maximum usefulness out of service personnel like soldiers and firemen, attacked the problem with renewed vigor. During World War II, the Air Force tried to use the new science of psychometrics to determine who would "endure the stresses of flying and combat," but had no luck, ultimately concluding that "the only valid test for endurance of combat is combat itself." In the 1980s, Canadian psychologist Stanley Rachman tried again, conducting a study of British Army bomb disposal experts. Despite running numerous tests on these men, he found it impossible to correlate their psychological profiles with their performance under pressure. It had begun to seem as though the mystery might remain unsolved.

In the last decade, though, technology has tilted the game in scientists' favor. Functional magnetic resonance imaging (fMRI) brain-scanning machines allow neuroscientists to peer directly into the activity of the living brain. These machines use strong magnetic fields to induce changes in the orientation of hydrogen atoms, which then respond differently to pulses of electromagnetic radiation depending on the chemical environment. Since active regions of the brain use more oxygen, an fMRI scanner can pinpoint which areas are busiest at a given moment with a lag of just a few seconds. These areas of activation are correlated with anatomical understanding generated through past research, such as brain-lesion studies, which have already provided a rough atlas of what functions are carried out where. With fMRI, then, researchers can see how different functional areas of the brain work together, without any danger or discomfort to the test subject.

Dr. Lilliane R. Mujica-Parodi thinks she can use fMRI technology to crack the mystery of fear. As director of the Laboratory for the Study of Emotion and Cognition at the State University of New York at Stony Brook, on Long Island, she wants to find out if there's a way to test someone in a normal setting that will identify how they will behave in a perilous one. She wants, in other words, to overturn Rochefoucauld's dictum.

What struck me about Mujica-Parodi's work, when I first came across it, was the unusual lengths to which she goes in order to elicit a vigorous fear response. For the last few years, Mujica-Parodi and her team have been examining human test subjects not only in the laboratory but also in the field. First, her subjects are hospitalized for several days, during which time they're asked to run on a treadmill and to provide blood, saliva, and urine samples. Then they're exposed to mild stressors—for instance, they're shown pictures of emotionally upsetting crime scenes. Finally, their brains are scanned in an fMRI machine.

So far, none of this is particularly extraordinary. Neuroscientists have been performing this kind of experiment for years. "Researchers have already shown that it's possible to predict with some efficacy, based on fMRI experiments, how people behave in response to a mild laboratory-induced stressor," Mujica-Parodi told me, when I called her on the phone to ask her about her work. "What we want to do is take it one step further and see if these measures that we see in the fMRI are actually predictive of a real-world stressor."

By "real-world stressor," what she meant was an experience that's intensely, grippingly frightening. For researchers, finding a way to explore how such a thing affects real people poses something of a challenge, as it's considered unethical to put people in actual mortal danger for the sake of psychological research. So Mujica-Parodi has come up with the next best thing. She recruits people who, of their own free will, have decided to go for their first-ever parachute jump at Skydive Long Island, a small operation about twenty miles east of the university. She does this by putting up a sign on the wall inside the jump center's office next to the airfield. The sign explains what she's doing and offers a small fee for volunteers. Anyone who takes her up on her offer is fair game, by ethics-review-board standards, for extreme fear research. And so Mujica-Parodi can run them through her tests, and compare their response to mild, laboratory-induced stressors with their physiological reaction to the sheer terror of freefall from twelve thousand feet.

As she told me this over the phone, I realized that I had stumbled on a rare opportunity: There couldn't be a better way to learn about cutting-edge fear research than to actually take part.

A FEW WEEKS LATER, I check myself into the research wing of Stony Brook Hospital for the first of the two-part experiment—the low stress-level part. I'll be at the hospital for two days, eating hospital food and sleeping on an adjustable hospital bed equipped with automatically inflating pouches in the mattress, the kind designed to prevent bedsores. It's important for me to stay in a controlled environment, Mujica-Parodi explains, so that her team can get an accurate measure of my body's baseline—a sense of what my hormones, heart rate, and other measures of stress are like when I'm in resting mode. I wear electrodes that measure my heart rate, and periodically the staff collects blood and saliva samples.

On the final morning, I lie down on a sliding table and am rolled inside an fMRI scanner, a donut-shaped machine that throbs and hums as its magnetic fields invisibly probe my brain. I remain motionless for forty-five minutes, staring up at a screen that displays pictures of different faces. Some are angry, others laughing, others wear blank or neutral expressions. It's my response to these latter faces that Mujica-Parodi is most interested in.

As I lie there looking at these faces, I'm not aware that I'm feeling any emotion one way or another. They're just pictures of faces. But deep inside my three pounds of gray matter, my fear centers are doing their thing. While I'm lying in the scanner, Mujica-Parodi is sitting inside a nearby control booth with a technician, looking at a cross-section of my brain on a computer monitor.

Later, she shows me some of the images. The most highly activated regions are lit up in red, orange, and yellow. The first region she points out is the thalamus, located in the middle of my head, at the top of my brain stem. A pair of bulb-shaped lobes, the thalamus operates as a kind of routing center in the brain's information superhighway, taking sensory data from the eyes and ears and body and shunting it off to the various parts of the brain. When I looked at the pictures of the faces, the light entered my eye, triggering neurons to fire in my retina, which sent a flow of information through the optic nerve to the thalamus. There the data stream split in two. What happened in the course of these two pathways lay at the heart of the mystery of fear.

The first route led to the amygdala, a pair of almond-shaped nerve centers located just in front of the thalamus and on either side. The second route headed to the neocortex, the deeply grooved and wrinkled tissue that lies on the surface of the brain. Specifically, it wound up at an area that lies just below the forehead, the frontal cortex.

I tell her that this is a region that I'm already familiar with. The frontal cortex is the part of the brain associated with the higher mental functions. It's a newcomer, in evolutionary terms. The main difference between human beings and closely related animals like chimpanzees and gorillas is that, in the last few million years, our frontal cortex has grown significantly bigger. It's not too much of an oversimplification to say that many of the higher faculties that we think of as uniquely human—our ability to reason and to plan ahead—are centered in the frontal cortex.

The amygdala, though, is less familiar to me. Mujica-Parodi explains that it's part of a much older region of a brain called the limbic system, which lies deep below the neocortex and handles many of the brain's more primitive processes, including emotion and memory. In particular, the amygdala is the key center for evaluating threat. As the visual information flows in from the thalamus, the amygdala scans it for any signs of danger.

The amygdala is also responsible for learning emotional associations. If I'm walking along the street and a dog bites my leg, I'll not only form a conscious memory of the event, but my amygdala will form a separate, subconscious one. Since I have no awareness of what memories are stored by the amygdala, if I don't have a conscious memory of an event, I might later find myself having an emotional reaction to something and not understanding why. For example, as infants, we're able to form long-term emotional memories long before we can form conscious ones. Leading fear researcher Joseph LeDoux of New York University has proposed that some phobias might arise when a person experiences something traumatic as a young child—a bee sting, say—that remains potent in the amygdala and thus able to inspire fear even without a conscious awareness of the cause.

When the amygdala identifies something in the environment as a potential threat, it triggers a series of alarms that result in all the outward signs of fear: trembling, sweating, blanching, and so forth. It also activates the insula, a region midway between the limbic system and the neocortex that's responsible for the conscious sensation of fear.

The amygdala sends numerous projections to the frontal cortex, and vice versa. From Mujica-Parodi's perspective, it's the nature of the dynamic between these two regions that's essential in understanding how a person responds to stress. If the amygdala detects a potential threat but the frontal cortex analyzes the data in more detail and determines that no threat is present, the cortex will tell the amygdala to quiet down.

Mujica-Parodi chooses a scan of my brain that was taken while I was looking at a neutral face, and points to a subregion of the frontal cortex called the ventromedial prefrontal cortex, or vmPFC. "The amygdala always responds to novelty," she explains. "When I show you a neutral face, the amygdala says: 'Oops, what's that: Is it dangerous?' And then the inhibitory component"—the frontal cortex—"kicks in and says, 'You know what, it's not. Calm down.'"

Because the faces I looked at weren't scary, my frontal cortex shut down the amygdala's response before I could become aware of it. The entire response was taking place automatically, outside of my consciousness. As far as I knew, nothing was happening at all. I was simply remaining calm. But while all this was going on, Mujica-Parodi was measuring the interactions between the two and testing the dynamic between the amygdala and the frontal cortex. From her analysis, she believes that she can predict how a person—me, in this case—will respond when they're well and truly terrified.

Now, to test this hypothesis, we have to move on to phase two. Real terror.

ON A CRISP CLEAR MORNING in early October I wake up before dawn in a motel room in eastern Long Island; shower, pack up my bags, and check out. I've slept poorly, troubled by uneasy thoughts that kept my mind skittering over the surface of sleep. As I drive along the Long Island Expressway I not only feel tired but also vaguely ill.

I drive past the abandoned Northrop Grumman factory where the F-14s featured in *Top Gun* once were built. It's just then, as I make a left-hand turn, that my anxiety spikes, and I wonder if I am going to make it through the morning. An awful sense of helplessness sweeps over me. I feel woozy and short of breath, but I decide to drive on, and tell myself that if I feel substantially worse, I'll stop. That idea calms me down a bit, and soon I'm at the edge of the airfield, where a small white building houses the office of Skydive Long Island.

To get the full measure of a test subject's response, Mujica-Parodi's assistants conduct a battery of tests before, during, and after each sky-diving session. As soon as I arrive, two graduate students in white lab coats usher me inside a trailer, where a burly phlebotomist extracts several vials of blood. I also give urine and saliva samples. Next, I strip down to my waist so that the research assistants can fasten electrodes to my chest. Then they help me into a special vest that will measure my lung volume. A satchel contains an altimeter to show when my jump starts and a digital recorder to carry all the data. Over that I don my shirt, a fleece, and a flight suit.

Then I'm led into another room in which I watch a mandatory video explaining the basics of tandem skydiving. It reiterates force-fully several times that, due to the multiply redundant waiver I've just signed, there's absolutely no way my heirs will be able to sue Skydive Long Island, regardless of the circumstances.

As I watch I feel acutely conscious of the gear strapped to my body. It's intended to monitor, I know, not the activity of my brain directly, but rather the effects of one particular subsystem, the autonomic ner-vous system (ANS), which is responsible for producing the symptoms of arousal in the body.

In evolutionary terms, the ANS is an ancient structure, even older than the amygdala. It dates back to a time when our ancestors were no more than primitive fish swimming in the Paleozoic sea. Back in those long-ago times, our ancestors' main concerns were fairly simple: eating, reproducing, and not being eaten. Evolution had furnished them with a nervous system that was commensurately simple; really nothing more than a network of neurons stretched out along the spinal column. Today, that's what our ANS still looks like: a ladder of knot-like nerve centers, or ganglia, stretching along the spine.

This system is divided into two opposing and complementary parts, the parasympathetic and the sympathetic nervous systems. When conditions are benign, the parasympathetic nervous system, PNS, comes to the fore, sending out impulses from the ganglia to organs throughout the body, encouraging them to do the sorts of things that one does when it's summertime and the living is easy. We relax. The heart rate slows down. Glands release saliva in the mouth and bile in the gut to assist in digestion. In the right context, the PNS helps generate an erection.

When danger rears its head, however, it's time for the parasympathic's counterpart to swing into action. The sympathetic nervous system, SNS, produces the famous fight-or-flight response, which prepares the body for the most basic kind of physical danger. If you're being attacked by a predator, the sympathetic nervous system instantly focuses all the body's resources on the priority of physical struggle. The adrenal medulla dumps adrenaline into the bloodstream. The heart starts pounding, delivering blood to the muscles where it's needed. Sweat glands open. Pain-deadening chemicals flood the brain. Digestion no longer being a priority, the mouth goes dry, and the peristaltic movement of the gut stops.

As I watch the video of people jumping out of an airplane over the Long Island countryside, my SNS jacks up another notch.

And then it's time to go. Duncan, the jump master, hands me a pair of gloves, a leather cap, and a pair of clear plastic goggles. He quickly reviews with me the procedure for the skydive, and then we climb into the back of the Cessna with a half-dozen other divers. As the pilot starts the engine and taxis to the end of the runway, I keep telling myself that I shouldn't be worrying, but it's no use. I'm suffering from all the symptoms of SNS activation: My heart's pounding; my mouth is dry; my stomach is churning. I feel ill. I try to concentrate on my breathing as the Cessna climbs steeply. *I can do this*, I tell myself. *I can stay in control.*

Duncan hands me a laptop so that I can do further tests. On the screen appear two checkerboard patterns; I have to compare them and decide if they're the same or different. The test is designed to gauge how effectively I can reason when fear is shutting down my ability to think logically.

Next, Duncan gives me a plastic test tube with a chunk of fibrous material in it that looks like the inside of a cigarette filter. I pop it in my mouth and chew it for thirty seconds. Later, lab technicians will measure the level of cortisol in my saliva, which will indicate to what extent my amygdala has activated another major system that the body uses to respond to danger: the hypothalamic-pituitary-adrenal, or HPA, axis.

Whereas the sympathetic nervous system is activated for all kinds of arousal, good and bad (e.g., fear, anger, and sexual arousal), the HPA axis is specifically about preparing the body for bad times. It's the heart of the stress response. In the face of danger, the amygdala sends a signal to the hypothalamus, the brain's gateway to the endocrine system. Unlike the nervous system, which controls the body through a network of neurons that carry electrical impulses, the endocrine system exerts control by releasing chemicals—hormones—into the bloodstream. When the body is subjected to stress, the hypothalamus sends a hormone signal to the pituitary gland, which sends another hormone to the adrenal gland, which then floods the bloodstream with the granddaddy of stress hormones: cortisol.

Like nuclear fission, cortisol can be used for good or evil. When the body needs to respond to an immediate threat, it's essential for allowing the body to metabolize emergency energy supplies. As cortisol surges through the body, it causes the release of energy-bearing sugars into the bloodstream and increases blood pressure, so as to more efficiently deliver nutrients and oxygen where they're needed. But when the body is subjected to lower levels of stress for a long time, cortisol has a cumulatively toxic effect, corroding the body's self-maintenance systems, leading to high blood pressure, decreasing cognitive functioning, and suppressing the immune system.

I'm feeling pretty toxic myself as the plane levels off and the door slides open to the rushing wind. Rationally, I know—that is to say, my frontal cortex is able to process the logical truth—that I'm not in any danger. But there's no way to convince my amygdala of that. That twelve-thousand feet of empty air is a reality that it understands perfectly well. There's simply nothing my frontal cortex can do. I'm being overwhelmed by the parallel mind of fear, as the amygdala floods my mind with thought-crushing molten lava of anxiety and dread.

The other skydivers jump; we scoot to the edge of the door. "Head back—ready—set—arch!" Duncan says, and out we go. We lean forward, and slip from the doorway into sudden weightlessness.

And then—nothing. My mind is blank, completely overwhelmed. Then, five or ten seconds later, I'm aware of tumbling, sky and earth changing places again and again. High up, a glimpse of the airplane. Then, quickly, we settle into a stable position, stomachs down and backs arched, like a double human shuttlecock. The fear is gone. Although my heart is pounding and the world still seems crisp and hyperreal, there is only an intense pleasure of thrill. Falling at terminal velocity, the wind howling around us, we no longer feel weightless. It's like we're lying on a cushion of air. There's no sense of speed at all. We could be floating motionless or flying like an airplane. I can see for miles. It's beautiful!

After about a minute, Duncan pulls the ripcord and we seem to be yanked violently upward. The rest of the descent is leisurely, almost pleasant, though my nerves are still humming like high-voltage wires.

Five minutes later, my feet touch down. The burly phlebotomist steps up bearing needles. "We're going to take some blood now." I don't mind at all. In fact, I'm elated. It's great to be alive.

AFTER I LEAVE, the research assistants return to Mujica-Parodi's lab at SUNY Stony Brook with the digital recorder containing my data. I wait for her to process the results, then I call her up to ask about what she's found.

Mujica-Parodi starts off by reminding me that she thinks that the secret to handling intense fear lies not in how forcefully the cortex and the amygdala struggle against one another, but in how their responses are coordinated. A vigorous amygdala response is crucial, she points out, in a world filled with dangers. Laboratory rats normally flee open spaces, but when they've had their amygdala removed they'll wander aimlessly in the open. In the wild, that's a good way to get eaten by a predator.

But it's just as dangerous to have an amygdala that's too active. When the fear system cranks up too high, performance starts to deteriorate. The amygdala's projections to the frontal cortex shut down organized thought, resulting in "brain freeze." Fine motor control disappears, hearing fades, and vision narrows into a tunnel. Any further, and you're in the throes of panic.

To maintain the balance between too little fear and too much, we rely on the amygdala and the prefrontal cortex to balance each other out. "The brain works as a control system circuit, like a thermostat in your home, with a negative feedback loop," Mujica-Parodi explains. The amygdala represents the excitatory component, and the prefrontal cortex one of the inhibitory components.

When dealing with fear, the goal is to bring your fear up to the point of maximum efficacy and no higher—to maintain control of your passion without letting it overwhelm you.

With that, Mujica-Parodi e-mails me a couple of tables and charts, and, once received, I call her back to discuss what they mean. One chart shows my heart rate as measured before, during, and after the jump, as well as another statistic, a figure labeled "sympathetic dominance." While this number roughly tracks my heart rate, it measures more specifically the arousal of my sympathetic nervous system.

When I was at rest, my baseline number was 4.9, which is about the same as most people. When I was in freefall, my number was 13.37, about twice the average. In other words, my mind was blown.

"Are you the sort of person who gets excited a lot?" she asks.

"I've been described as kind of hyper," I admit.

"From what I've seen here, that's also the way I would describe it."

But what was more meaningful than my response during freefall was my condition before and after it. Riding up in the plane before the jump, my sympathetic dominance was rated at 5, significantly less than the average of 6.5. After the jump, it quickly fell to 8.4.

"What this means is that you have good selectivity," Mujica-Parodi said. "You're kind of like a yo-yo, in that you're conserving your sympathetic dominance for when it's actually needed. You go up, and you come back down right away."

These results, Mujica-Parodi tells me, mirror those of my fMRI session: My negative feedback loop is tightly coupled. This, in her view, is the essence of stress resilience. It's not that I stay cool when I'm plummeting toward earth—"You were in actual danger," she says, so "a strong excitatory response is appropriate"—but that when I'm not falling I'm suppressing the fear response and conserving my energy.

If Mujica-Parodi is right, then she's found the holy grail of stress research: a way to predict how a person will perform under high-stress situations by examining them beforehand in the calm of a laboratory. Such indicators are of obvious interest to the military, which is why the Navy is backing her research. It could also prove useful in recruiting and training police officers, firefighters, and anyone else who might face danger in the line of duty. The technology could even be used to diagnose mental disorders, such as schizophrenia, which seems to involve very poor coupling of the amygdala and cortex.

Ultimately, her research could help the rest of us, too. By explaining how its machinery really works inside our brain, we could all learn to better deal with extreme fear. After all, danger can reach into anyone's life, at any time. If we can learn to understand the strange parallel mind of fear—the subconscious center that's forever plugging along out of sight, assessing what's going on, making its own decisions, and triggering its own responses—we might be able to deal with it better when it takes over.

Otherwise, we'll just have to hope for the best. Like Tom Boyle, Jr., did one warm summer evening in Phoenix.

CHAPTER TWO

SUPERHUMAN

HERE'S HOW IT IS: one minute, you're going through your daily routine, only half-paying attention. And the next you're sucked into a vivid, intense world, a place in which time seems to move slower, colors seem brighter, and sound seems more perceptible, as though the whole universe has suddenly come into focus.

It was about 8:30 p.m. on a warm summer evening in Tucson, Arizona. Tom Boyle, Jr., was sitting in the passenger's seat of his pickup truck, his wife Elizabeth at the wheel, waiting to pull out into traffic from the shopping mall where they'd just had dinner. They were in a good mood. The kids were at the grandparents', and they'd been able to enjoy dinner together, just the two of them, chatting and giggling, for the first time in a long while.

The Chevy Camaro ahead of them hit the gas, spun his wheels, and jerked out onto the avenue with a squeal of rubber. "Oh my God," Elizabeth said. "Do you see that?"

Boyle glanced up to see a shower of sparks flying up from beneath the chassis of the Camaro. And something else: a bike, folded up from impact. And a rider. The Camaro had hit a biker, and the rider was pinned underneath the car. Without thinking, Boyle threw open the door of the truck and started running after the car, his heart in his throat.

For a few gruesome seconds, the Camaro plunged on, dragging along the rider, eighteen-year-old Kyle Holtrust. One of Holtrust's legs was pinned between the chassis of the car and the frame of his bike; the other was jammed between the bike and the asphalt. After twenty or thirty feet, the Camaro slowed and stopped. The rider and bicycle were a tangled mass lodged underneath the front of the car. Holtrust was screaming in agony, howling like a wounded animal, pounding on the side of the car with his free hand.

Boyle didn't even think about it: just reached under the frame of the car and lifted. With a sound of groaning metal, the chassis eased upward a few inches. "Mister! Mister! Higher! Higher!" Holtrust screamed.

Boyle braced himself, took a deep breath, and heaved. The front end lifted a few more inches. "OK, it's off me!" the boy called out, his voice tight with pain. "But I can't move. Get me out!"

The driver of the car, forty-year-old John Baggett, was still sitting behind the wheel, looking numb. "Hey buddy! Hey!" Boyle shouted. "Get out here! Pull the kid out!" But Baggett just sat there numbly. "Hey, asshole!" Only after Boyle had yelled at him four or five times did Baggett snap out of his stupor, reach down, and pull Holtrust free. At last, about forty-five seconds after he'd first heaved the car upward, Boyle set it back down.

The biker was badly hurt, in a lot of pain, and frightened. Blood was pouring out of his wounds. Boyle knelt down and wrapped the young man in his arms, comforting him until the police and fire department arrived. Holtrust seemed strangely composed, thanking Boyle for his help and expressing regret at having inconvenienced everyone.

After the ambulance took Holtrust away, Boyle collapsed onto the ground. "Hon," he said to his wife, "I've got to get home. I feel like I'm going to throw up."

The local media celebrated Boyle's feat of compassion. The YMCA gave him an award. Newspapers and TV stations interviewed him. The fanfare flattered him and he felt extremely proud of himself.

Yet to this day there's something about that evening that he can't figure out. It's no mystery to him *why* he did what he did—"I would be such a horrible human being if I could watch someone suffer like that and not even try to help," he says—but he can't quite figure out *how*.

"There's no way I could lift that car right now," he says.

Boyle, it should be pointed out, is no pantywaist. He carries 280 pounds on a 6′4″ frame. But think about this: The heaviest dumbbell that Boyle ever deadlifted weighed 700 pounds. The world record is 1,003 pounds. A stock Camaro weighs 3,000 pounds. Even factoring leverage, something extraordinary was going on that night.

That something was the body's fear response. When we find ourselves under intense pressure, fear unleashes reserves of energy that normally remain inaccessible. With the sympathetic nervous system and the hypothalamic-pituitary-adrenal (HPA) axis fully activated, our bodies and brains can utilize their resources so fully that we become, in effect, superhuman.

The positive effects of fear have been recognized for thousands of years, but it was not until the beginning of the twentieth century that the benefits of acute stress came under scientific scrutiny. In 1908, American psychologists Robert M. Yerkes and J. D. Dodson published a seminal paper that laid the groundwork for stress research. The two men trained mice to choose between two doors: one white and the other black. If a mouse chose the white door, it was allowed to pass through a corridor that led back to their nesting area. If it chose the black door, it received an electrical shock and had to return to its nesting area through the white door. Yerkes and Dodson found that the stronger the shock, the more quickly the mice learned to choose the correct door—at least if the shock was mild or moderate. But if they gave the mice a very strong shock, their ability to learn deteriorated. They seemed to be breaking down under pressure.

Before long, other researchers had extended Yerkes and Dodson's findings to a much more general understanding of performance under stress. The Yerkes-Dodson law states that increased stress leads to increased performance up to a certain level of intensity, beyond which performance gradually levels off and then declines. Because of the shape of the graph described, this idea is sometimes referred to as the "inverted U" model of stress reactivity.

It makes sense, intuitively, that a low to moderate degree of jitters can improve performance. If you go out for a jog around the park on Sunday morning, your arousal level will be low. Even if you try to run as fast as you can, you won't put all your effort into it. But if there's something at stake—say, you're running against the crowd at the annual 5K race—then your arousal level will be higher. You'll feel focused and motivated. Chances are you'll shave a few minutes off your usual Sunday-morning time.

Researchers have found a correlation for this effect in the neurochemistry of the brain. As we become increasingly stressed, the amygdala stimulates a region of the brain stem called the locus ceruleus. This area is the brain's main factory for the production of noradrenaline, the form of adrenaline used within the brain, and sends out direct connections to the prefrontal cortex, the area of the brain associated with higher cognitive function. Dosed with noradrenaline, the prefrontal cortex becomes more active and focused. Related compounds have similarly stimulating effects elsewhere in the body.

Primed in this way by fear, we feel alert and empowered. And it's not just a feeling. Thanks to the sympathetic nervous system, we're capable of performance far beyond our normal limits.

SPEED

Fear gives us speed—both in how quickly we can respond to a threat and in how fast we can move our bodies. In a world filled with all kinds of danger, a rapid response is essential to survival. While some threats can unfold over minutes, hours, days, or years, others—a lion jumping out from behind a bush, a landslide roaring down a hillside—can be upon you in an instant. A slow response can mean the end.

To deal with threats occurring on varying time scales, we have a multilayered defense system, with different alarms going off to alert us to danger at different time scales. The fastest alarms are the crudest and least discriminating. Each successive layer is slower but also finer-grained and more sophisticated in its ability to discriminate between threat and false alarm.

The brain's first line of defense is the startle reflex, a lightning-fast but undirected reaction to potentially hazardous changes in the environment. When you hear a loud "*bang!*" for instance, the sonic vibrations in the air trigger a neuronal firing inside your ear. Within five milliseconds, an extremely simple chain reaction involving just three neurons in the brain stem and spinal cord has triggered the startle reflex. In an instant, hundreds of muscles all over the body are recruited into an integrated, synchronized spasm: your eyes clamp shut, your neck and shoulders and chest tighten up, your abdominals harden, your hands clench, and your elbows stick out. Your pupils contract and your heart rate shoots up. The reflex is an ancient one, in evolutionary terms, and in the distant past it probably helped save our ancestors from sudden danger by bracing them for impact.

The more afraid you are, the more easily the startle reflex is triggered. The phenomenon is known as "fear-potentiated startle." Some people call it "being jumpy as a cat." If you watch a scary movie alone at home on a dark night, then turn off the lights to go to bed, you know that the slightest sound will make you leap. That's the effect of fear-potentiated startle. These days, researchers use the effect to work backward, to figure out how stressed a lab animal is by measuring how easily it startles.

As the startle reflex is taking place, information is flowing through the thalamus and splitting into two streams, one headed for the amygdala and the other for the cortex. The latter route is much quicker: It takes just twelve thousandths of a second for a signal to reach the amygdala from the ear, half the time it takes to reach the cortex. The amygdala processes data more quickly, too, giving it a quick once-over for similarities to previously identified hazards and lighting up at the slightest suggestion of danger. It's not particularly fussy or detailed in its analysis. As we've seen, a face doesn't have to be angry or frightened for the amygdala to light up; merely being unfamiliar is sufficient.

The slowest part of the alarm system is the cortex. Not until this region of the brain processes the signal does it enter our consciousness, a process that takes at least half a second—an eternity in the timescale of reflexes. Generally, the cortex acts to inhibit the faster, cruder alarm systems by identifying false alarms and shutting off the warning bells. This is what Lilly Mujica-Parodi was observing as she watched my ventromedial prefrontal cortex turn off my amygdala when I looked at faces in the functional magnetic resonance imaging (fMRI) scanner.

The salient point to remember about the cortex is that it's slow. By the time we even become conscious of a threat, the amygdala might already have taken steps to deal with it. Indeed, the phenomenal speed of the brain's fear response is such that we can find ourselves acting so quickly that it feels like we're doing it before we even realize we want to. When fear is strong, the amygdala can trigger automatic patterns of behavior without waiting for involvement by the conscious mind. These automatic behaviors are essentially reflexes that we can learn and store. As we go about our day, a good portion of what we do consists of motor patterns that we carry out more or less unthinkingly. It's why most of us can walk and chew gum at the same time: we don't have to think about either one, because both have been stored as simple motor patterns.

The more you practice a motor pattern, the quicker and more effortless its execution becomes. Eventually, your brain can deploy it automatically without any conscious thought at all. Rob Smithee★, a Green Beret, told me of an incident during his service in Mogadishu when he saw, out of the corner of his eye, a Somali running at him with a hatchet. Before he fully realized what was happening, he saw a dark blur in the periphery of his vision: his own pistol, rising toward the attacker.

"My pistol was up, out, and I shot him before I even knew it," Smithee recalls. "I fired four times, and hit him with three bullets."

That kind of response, Smithee adds, took a great deal of preparation to establish as an automatic motor pattern. "I'd fired literally thousands of rounds on a firing range," he says.

FOCUS

As Samuel Johnson observed, "When a man knows he is to be hanged in a fortnight, it concentrates his mind wonderfully." But while mortal peril of every kind is notably effective at clearing out one's mental clutter, even at less intense levels fear conveys impressive powers of focus.

"Normally your brain is doing lots of things at the same time, just under the surface of awareness," says David Eagleman, a neuroscientist at the Baylor College of Medicine. "Your brain is thinking about where you're going to eat lunch, and what you're going to wear to the party tonight, and what you're doing for your career, and so on. What happens in a really scary situation is that the amygdala essentially tells the rest of the brain, 'Hey, everybody shut up and pay attention to this.' All the nonessential processes get shut down and your whole brain, or as much of your processing power as you have, gets devoted to this one thing going on. And as a result, one of the things that happens when you're afraid is that you can think very clearly about the situation."

The effect is due, at least in part, to the effect of noradrenaline within the brain. At low to moderate concentrations, a type of protein within the cell walls of neurons called the α2-neuroadrenaline receptor responds to the release of noradrenaline and acts to increase neural efficiency. One of its effects is to boost activity in a brain center called the right ventral lateral prefrontal cortex. This region is central to the cognitive control of thoughts and emotions. It helps keep us on target. With the noradrenergic pedal to the metal, we're vigilant, we're present, and we can think clearly.

That's why fear can be an invaluable tool in the workplace. It's easy to let paperwork slide right up until a deadline. Only then, with the threat of disaster looming, can we forget about all the clutter and distractions around us and finally focus on what needs to be done.

For athletes, achieving a state of optimal arousal can feel transcendent. Skiers, big wave surfers, and others talk about finding themselves in "the Zone," a place of intense focus and total concentration in which, paradoxically, one feels calm and at peace. Dean Potter, a thirty-five-year-old rock climber and tightrope walker, recently described for a *New York Times* reporter how he enjoyed the sensation of balancing hundreds of feet above a rocky canyon without a safety harness. "When there's a death consequence, when you are doing things that if you mess up you die, I like the way it causes my senses to peak," he said. "I can see more clearly. You can think much faster. You hear at a different level. Your foot contact on the line is accentuated. Your sense of balance is heightened. I don't seem to feel that very often meditating."

Long after Neil Williams survived his crash landing in a Zlin, the sights and sounds he experienced that day remained vivid. As he wrote in a memoir: "I sat on the grass and realised how all my senses had been heightened by the drama of the last few minutes: the colour of the grass and sky, the smell of the earth, the song of birds: never before or since has it been so clear."

MEMORY

It's telling that Williams remembered so many vivid details of what happened during his near-catastrophe in the Zlin years after the fact. The same jolt of noradrenaline that spurs the cortex to greater mental focus amps up the hippocampus, the region near the amygdala that stores explicit memories.

In an experiment reminiscent of the famous Yerkes–Dodson protocol, psychologist Christa McIntyre of the University of California, Irvine, let rats walk into either a dark or a well-lit chamber. Being nocturnal creatures, most chose the dark chamber, where they received an electrical shock. The shock wasn't very strong, and apparently didn't make much of an impression; rats put in the same situation a day later went right back to the dark chamber. When the rat's amygdala were chemically stimulated before they were shocked, however, the story was different. This time they remembered the shock so vividly that they shunned the dark room and preferred the light one. "Emotionally arousing events tend to be well-remembered after a single experience," says McIntyre, "because they activate the amygdala."

Strange to say, but while most of us try to avoid stress in the course of our daily lives, it's the stressful, emotionally intense memories that will live with us the longest. Perhaps that's why we keep returning to high-school and college reunions. Unshielded by experience, adolescents feel pain in a way that middle-aged people never do, an amygdala-twisting cavalcade of angst, love, heartbreak, love, excitement, and despair. But they form indelible memories. Before she died, my grandmother lived in a nursing home with other men and women in their eighties and nineties. Some of them couldn't tell you what year it was, but their memories of World War II, when they'd been young and alive and frightened, were as clear as ever.

It makes sense, from an evolutionary point of view, that we need to remember most vividly the events of high-pressure situations. The "flashbulb memory effect" is nature's way of making sure that we know what to do if we're in a similar situation again.

These things not only stay in our memory longer, we're able to recall them in greater detail. Research suggests that we're not storing more information about the event as it happens, but rather that the hippocampus is holding onto more of our impressions as they unfold and keeping them longer. Smaller events that might have been discarded almost instantaneously instead get put in the "keep" folder, so that looking back it seems like much more happened.

This richness of emotionally charged memory led David Eagleman, the neuroscientist at the Baylor College of Medicine, to wonder what was behind another psychological effect of fear: time dilation, or the apparent slowing-down of time. It's a common trope in movies and TV shows, like the memorable scene from *The Matrix* in which time slows down so dramatically that bullets fired at the hero seem to move at a walking pace. Eagleman reasoned that when time seems to slow down in real life, our senses and cognition must somehow speed up—either that, or time dilation is merely an illusion. This is the riddle he set out to solve. "Does the experience of slow motion really happen," Eagleman says, "or does it only seem to have happened in retrospect?"

To find out, he first needed a way to generate fear of sufficient intensity in his experimental subjects. Instead of skydiving, he found a thrill ride near the university campus called Suspended Catch Air Device (SCAD), an open-air tower from which participants are dropped, upside down, into a net 150 feet below. There are no harnesses, no safety lines. Riders plummet in free fall for three seconds, and then hit the net at seventy miles per hour.

Was it scary enough to generate a sense of time dilation? To see, Eagleman asked subjects who'd already taken the plunge to estimate how long it took them to fall, using a stopwatch to tick off what they felt to be an equivalent amount of time. Then he asked them to watch someone else fall and then estimate the elapsed time for the plunge in the same way. On average, participants felt that their own experience had taken 36 percent longer. Time dilation was in effect.

Next, Eagleman outfitted his test subjects with a special device that he and his students had constructed. They called it the perceptual chronometer. It's a simple numeric display that straps to a user's wrist, with a knob on the side to let the researchers adjust the rate at which the numbers flash. The idea was to dial up the speed of the flashing until it was just a bit too quick for the subject to read while looking at it in a nonstressed mental state. Eagleman reasoned that, if fear really does speed up our rate of perception, then once his subjects were in the terror of freefall, they should be able to make out the numbers on the display.

As it turned out, they couldn't. That suggests that fear does not actually speed up our rate of perception or mental processing. Instead, it allows us to remember what we do experience in greater detail. Since our perception of time is based on the number of things we remember, fearful experiences thus seem to unfold more slowly.

Eagleman's findings are important not just for understanding the experience of fear, but the very nature of consciousness. After all, the test subjects who fell from the SCAD tower certainly believed, as they accelerated toward freefall, that they knew what the experience was like at that very moment. They thought that time seemed to be moving slowly. Yet Eagleman's findings suggest that that sensation could only have been superimposed after the fact. The implication is that we don't really have a direct experience of what we're feeling "right now," but only a memory—an unreliable memory—of what we thought it felt like some seconds or milliseconds ago. The vivid present tense we all think we inhabit might itself be a retroactive illusion.

STRENGTH

As Tom Boyle found out through firsthand experience, fear (whether for ourselves or for someone else) can make us uncannily strong. Under acute stress, the primary job of the sympathetic nervous system is to prepare the body for sustained, vigorous action. As the adrenal gland dumps cortisol and adrenaline into the blood stream, blood pressure surges and the heart races, delivering oxygen and energy to the muscles. It's the biological equivalent of opening the throttle.

Vladimir Zatsiorsky, a professor of kinesiology at Penn State who has extensively studied the biomechanics of weight lifting, draws the distinction between the force that our muscles are able to theoretically apply, which he calls "absolute strength," and the maximum force that they can generate through the conscious exertion of will, which he calls "maximal strength." An ordinary person, he has found, can only summon about 65 percent of their absolute strength in a training session, while a trained weightlifter can exceed 80 percent.

Under conditions of competition, a trained athlete can improve as much as 12 percent above that figure. Zatsiorsky calls this higher level of performance "competitive maximum strength." This parameter is not a fixed number—the more intense the competition, the higher it can go, as the brain's fear centers progressively remove any restraint against performance.

It's no coincidence that world records in athletic events tend to get broken at major events like the Olympics, where the stakes are highest and the pressure is the greatest. Of the eight gold medals that Michael Phelps won at the 2008 Olympics, for instance, seven were world records. Not only that, but when he crossed the finish line in the men's 100-meter butterfly in 50.58 seconds, breaking the previous

Olympic record, three of the other seven swimmers who finished after him also came in ahead of the previous record.

But there's a limit to how fast and how strong fear can make us. We've all heard stories about panicked mothers lifting cars off their trapped babies. They've been circulating so widely, for so long, that a great many assume that they must be true. Zatsiorsky's work, however, suggests that while fear can indeed motivate us to approach more closely to our absolute power level than even the fiercest competition, there's simply no way to exceed it. A 100-pound woman who can lift 100 pounds at the gym might, according to Zatsiorsky, be able to lift 135 pounds in a frenzy of maternal fear. But she's not going to suddenly be able to lift a 3,000-pound car. Tom Boyle was an experienced weight lifter. The adrenaline of that June night gave him an edge, but it didn't turn him into the Incredible Hulk.

The mechanisms by which the brain is able to summon greater reserves of power have not been well explored, but they may be related to another of fear's superpowers: analgesia, or the inability to feel pain. When I'm at the gym, straining to complete the last repetition of a dumbbell exercise, it's pretty hard to imagine that my muscles have the capacity to work half again harder than they already are. What I feel is screaming agony.

But under intense pressure—whether it's a bodybuilding competition, a kid trapped under a car, or an attacking bear—you just won't feel that pain. The body pulls out all the stops and lets you turn the dial up to "11." You don't feel the ache of your muscles. You don't feel the pain. You just do what needs to be done.

The biochemical basis for fear's painkilling power has been well established experimentally. Under stress, the brain releases two groups of chemicals, endocannabinoids and opioids. As their name implies, these compounds have effects very similar to the certain illicit drugs—marijuana and heroin, respectively. Once released within the brain, they deaden pain.

Dave Boon, a motivational speaker who lives in Denver, got a taste of stress-induced analgesia in January 2007. He and his wife, June, were driving on Highway 40 west of Denver, heading for a weekend of skiing at Winter Park in the Colorado Rockies. Thousands of feet above them, the fan-like shapes of avalanche chutes stretched down the flanks of Stanley Mountain. The sky was cloudless blue. "Those chutes haven't run in quite a while," Boon remarked to his wife.

It must have been at almost that exact moment that, 2,200 feet above, the smooth white snowpack fractured. A minute later, a puff of powder danced across the highway, and an instant after that a 125 mph blast of wind slammed the car into the guardrail. "I can't see past the windshield, and everything's moving in slow motion," Boon remembers. "And my brain's going, 'What's going on? I don't understand what's going on here.'" Then the avalanche hit, T-boning the car with the force of hundreds of tons of snow, tree trunks, and debris.

"It blew us up and over the guardrail and we did a couple of flips in the air, and we hit and started to roll, and it just kind of went dark," Boon says. "I'd driven this road literally a thousand times over the last twenty years, and I knew how steep the slope was. About the second roll I thought to myself, 'We're going to go for a long ways.'"

Even though his car was being churned around inside an eighteen-foot-deep, roaring maelstrom of snow, rocks, and trees, Boon says, "It was quiet. We were going: '*thwump, thwump, thwump.*' Kind of like muffled tumbling."

Dave and June Boon were thrashed and pummeled as they rolled down the mountain, and a tree branch four inches in diameter punched through the car and slammed into Dave's arm, yet at the moment he felt nothing.

As the car came to rest upside down, the avalanche training they'd had twenty-five years before kicked in. June shouted: "Make an air space! Make an air space!" Dave cleared a tunnel out through the broken side window up to the surface, then returned to help June, who was pinned in her seat. As he cleared away the snow and broken glass from around her face, more snow kept falling down. June nearly panicked, but Dave managed to calm her, and with the help of a passerby managed to pull her free. Together they hiked uphill back to the highway, where an ambulance picked them up and took them to the hospital for treatment of their injuries. They were discharged that afternoon.

That night, as their stress-induced analgesia wore off, Boon and his wife found themselves in terrible pain. "When the avalanche hit, I actually popped a rib, and they didn't pick up on it in the hospital," Boon says. "I had a huge bruise where the four-inch diameter limb went right through our car and ricocheted off my right arm. We didn't sleep that night, we were so sore. We finally crawled out of bed at 7 a.m. the next morning, not having slept a wink."

GOING BERSERK

Great strength, speed, endurance, and the ability to withstand pain make for a formidable combination in an individual engaged in potentially fatal physical violence. The effects of acute fear, indeed, can be so impressive that others might find them hard to credit. Throughout history, legends have arisen about the preternaturally formidable abilities of certain classes of warriors.

According to the Norse sagas, for three centuries, from about 800 to 1100 AD, Scandinavia was plagued by roving bands of thugs and rogues called Berserkers. They took their name from the legendary Norse hero Berserk, a man of unequaled courage and fury in battle. Instead of armor in battle, he wore only *"ber sark,"* a bearskin. Like their namesake, the Berserkers upheld the principles of savage violence, and in combat wore no armor. Their signature activity was *berserksgang,* or going berserk, which involved roaming in bands and creating maximum destruction. According to the Norwegian historian Peter Andreas Munch, during these outings the Berserkers "were seized by a wild fury, which, at the moment, doubled their strength and made them insensible to bodily pain, but which also deadened their humanity and reason, and made them like wild animals.... Men who were thus seized performed things which otherwise seemed impossible for human power."

A thousand years later, the U.S. Army encountered a foe it found equally superhuman. Following the occupation of the Philippines during the Spanish-American war, the United States found itself unexpectedly bogged down in fighting a guerilla insurgency. Having no experience in jungle warfare, the Americans were unprepared for the hit-and-run tactics of the indigenous fighters, especially in the lush and mountainous island of Mindanao, home to the fierce Muslim tribes known as the Moro. Here patrols had to pick their way along narrow trails close-hemmed with seven-foot-tall elephant grass, where they were easy prey for ambushes by swarms of fighters armed with sharp, machete-like knives.

Amid the rush of the initial onslaught, a startled soldier might have time to get off only a single shot from his pistol. To their dismay, the American troops found too often that a single bullet wasn't enough against the fierce Moro. One junior officer wrote that "in hand-to-hand combat our soldiers are no match for the Moro. If our first shot misses the target, we rarely have time to get off another." Another officer reported that his soldiers had several times shot attacking Moro multiple times with a standard-issue revolver, only to be cut to pieces by their still-charging foe.

Amid a hubbub of similar complaints, the federal government contracted to the Remington Rand company to develop a semiautomatic pistol with greater stopping power. The result was the M1911, better known as the Colt .45. It became the standard-issue sidearm for the U.S. military for more than seventy years.

Many observers have had a hard time crediting that human beings could possess such extraordinary traits by solely natural means. During the time of the Philippine insurgency, the American press repeatedly described the Moro warriors as being high on pain-killing drugs. More sober research later found that this was not at all the case. Nor, for that matter, are we to credit speculation by twentieth-century historians that the Berserkers arrived at their state of wild-eyed frenzy thanks to the effects of the psychedelic mushroom, *Amanita muscaria*. Such explanations simply aren't necessary. Humans in the throes of a full-blown stress response don't need to take drugs. Terror is a powerful drug on its own. And here we hit upon one of the many paradoxes of fear: This awful emotion, which most of us do so much to avoid, can make us in many ways superior to our normal selves. Under its spell, we enjoy intense awareness, clarity of thought, and vivid memories. We're stronger, quicker, and have better endurance.

Wouldn't it be great if we could be like this without the unpleasantness of fear? Here's the thing: we can. Many of the illegal and illicit drugs that people take are efficacious exactly because they flood the brain with ersatz versions of the chemicals we produce naturally under stress. As we've seen, marijuana and opiate drugs like heroin mimic the pain-killing properties of compounds that are released naturally within the brain during stress. Amphetamines are structurally very similar to norepinephrine, and convey many of the same benefits in their action upon the central nervous system: improved reaction time, relief from fatigue, improved alertness, and better cognitive function.

No wonder these drugs are so habit-forming. It's very easy to form an uncontrollable craving for the benefits of fear when we don't have to endure the fear itself. Or even when we do. Thrill-seekers who habitually surf huge waves, jump out of airplanes, or hurl themselves from SCAD towers are all rewarded by the same surge of mind-altering chemicals that heroin addicts and speed freaks are after. The term "adrenaline junkies" really isn't too far off the mark.

WE'RE ALMOST finished with our tally of the superpowers that fear gives us. There's just one more: the ability to carry on as though we're not afraid at all. Survivors of disasters and other life-or-death crises often talk about how they found themselves simply doing what they needed to do, without panicking or indeed feeling any fear at all. Swaddled in a kind of emotional analgesia, they just carry on. Only later, when the effects of the fear response wear off, does the full emotional impact of what they've been through hit them.

Dave Boon recalled that, as his car turned over multiple times, he wasn't conscious of feeling fear. The thought that he might die never crossed his mind. Instead he found himself observing what was going on with dispassion. "When we were rolling, I thought to myself how steep that slope was, and I thought: *This is a massive avalanche, to be doing this to our car*," he says. "That's all I thought: *This is a massive avalanche*."

It wasn't until four days later, when Boon finally saw his car in the wrecker yard, that he realized the full impact of what he'd been through. "When I saw my car, my legs just buckled," he says. "I just sat down in the middle of the parking lot. I thought, *Holy crap, how did I get out of this?*"

He was lucky. For many who find themselves in mortal danger, fear provides no superhuman benefits. Instead, it triggers an immediate and catastrophic collapse. This is the dark side of fear.

CHAPTER THREE

LOSING IT

AS FEAR RISES above the levels that confer optimal performance, it enters a mysterious and much-feared territory: the dark side of the Yerkes-Dodson curve. Here, the magical upslope of enhanced abilities and superhuman performance starts to veer irretrievably downward. This is the realm we dread, the zone where we lose all the benefits that fear can confer and instead plunge toward incapacitation. We tremble, we sweat, we freeze. We find it hard to think. We're seized by the urge to scream or run away. We feel like we're losing control. When people talk about battling fear, this is the combat zone.

For psychologists studying the fear response, the degradations caused by extreme arousal are of immense interest. If they can learn how to mitigate them, they will have essentially tamed the fear response. One major hurdle, however, is that to study the phenomenon research-ers would first have to subject their human test subjects to intense terror—trials longer and more severe even than skydiving. And that's something that no ethics review board will permit.

In the 1950s, however, ethical guidelines were a good deal more relaxed. It was in that context that one of the most remarkable psycho-logical experiments ever conducted took place: a series of tests carried out by the U.S. Army called Task FIGHTER.

THE RECRUITS WERE young, most still in their teens. They'd only been in the army a few weeks, hadn't even finished basic training, but that was long enough to know that when you got an order, you didn't ask questions. So they packed up their gear and rode to the airfield. Once there, they were told that they would be boarding a military transport to take part in a psychological study about how altitude affects motor skills. As standard procedure, each man was issued a life preserver and a parachute. The prop wash from the DC-3's rumbling twin engines lashed the tarmac as they climbed up the folding stairs and filed into the cramped fuselage. An army psychologist and a male flight attendant boarded with them.

The year was 1962, and the Cold War was at its peak. The United States stood warily astride the world, an economic and military colossus. From Berlin to the Bering Straits, half the globe lay under the dominion of hostile communist countries, with nuclear capability and millions of men under arms. To prevail, the West would need all the advantages of technology in addition to its enormous wealth and political leverage. Scientists could build faster planes, longer-range rockets, and better ships. But could scientific research improve the quality of its fighting men?

The question couldn't have been further from the soldier's minds as they buckled into their seats. For many of them this was their first ride in an airplane. The California coastline slipped away as the plane climbed to five thousand feet and leveled off. The psychologist handed out test papers. Soon after the men had completed them, the plane lurched ominously.

A view out the window brought unsettling confirmation: One of the propellers had stopped turning. Over the intercom, the men could hear the pilot talking with the control tower about mechanical problems. Then the pilot addressed them directly: He was declaring an emergency and heading back to the field. The plane banked, and as it passed the airfield the men could see fire trucks and ambulances far below, racing to take position. Evidently a hard landing was in store.

As the men contemplated their fate, the pilot came on the intercom again. The situation had gotten worse. Now the landing gear was malfunctioning. The pilot ordered the flight attendant to prepare for an ocean ditching. Now, instead of being met by a rescue crew, the plane would come to rest in the deep, cold waters of the North Pacific. Hardly a man could fail to recognize that his chances of surviving had just dropped significantly.

In preparation for the impact, the steward began handing out paperwork. First came the Emergency Data Form, followed by an Emergency Instructions Test. The steward explained that, by answering the questions on the forms, the men would demonstrate to the insurance company that they had indeed received the required preflight safety briefing. Only with that paperwork completed, the steward said, could the men's surviving families be paid. Lest this information cause them any alarm, the steward assured them that before impact the forms would be put in a waterproof container and jettisoned into the ocean for later recovery.

The men were left alone with their pencils, each to deal with his fear in his own way. Some trembled, or sobbed, or stared forlornly. Due to the noise and the seating arrangement inside the airplane, it was impossible for them to confer with one another. They worked on their forms in isolation, alone with their thoughts.

And then the plane landed. They were back at the airfield. No explanation was given for this unexpected reprieve. The men were taken to a nearby classroom, where they were given another form, which asked them to describe their emotional state. Then they were told the truth: They had never been in danger. The ride had been part on an ongoing Army psychological experiment, whose goal, as the researchers later put it, was "to expose individuals experimentally to a hostile environment" in hopes of better understanding the "degradation of behavior in combat."

The forms had been psychological tests in disguise. As revealed by their scores, the men had indeed suffered a marked decline in mental acuity, committing three times as many mistakes on the Emergency Instructions Test as control subjects who'd stayed on the ground, their minds unfogged by mortal danger.

Other groups of recruits were put through other similarly harrowing scenarios. In one, men were sent one by one to a remote radar post and then subjected to an "accidental" artillery shelling. Dynamite exploded around them. In that scenario, the subjects had to follow complicated instructions to rewire a radio so that, ostensibly, they could call in a rescue helicopter to evacuate them. Not surprisingly, as the howitzer rounds exploded around them, few completed the do-it-yourself electronics project, and several tried to run away. Remarkably, according to the experimenters, after the men were told the truth of the situation, "in no case was there any evidence of residual tension or negative feelings."

Whether that's true or not, an experiment like Task FIGHTER would never be permitted today. Any research involving human beings is subject to strict oversight. Subjects may not be tormented without their prior consent—even if they later report no hard feelings. At any rate, the experimental results garnered as a result of its elaborate ruses were rather slim. Mostly, they confirmed the unsurprising fact that soldiers who are frightened, confused, and ill-led are very unlikely to perform well under intense stress.

DECIPHERING THE CURVE

Though their methods were irregular, the scientists of the Task FIGHTER project were in pursuit of a substantive goal that's still of interest to researchers today: How and why does performance decline when the brain is in the grip of extreme fear?

From an evolutionary perspective, it's strange that we should even have to worry about this problem. It doesn't make sense that our bodies are equipped with a system whose main function, when the going gets tough, is to incapacitate us. As we've seen, low and moderate levels of stress can spur performance benefits. These would have helped our ancestors survive and pass along their genes. It's easy to see why such mechanisms would have evolved. But a deterioration of performance under high stress makes less sense. It seems a setup for an evolutionary dead-end.

As they looked for an answer to this riddle, scientists soon realized that the situation was more complex than it had at first seemed. The essential idea of the Yerkes–Dodson curve—that performance follows a simple inverted-U pattern under increasing stress—was an oversimplification. There isn't one curve, but many. Some kinds of tasks are more affected by stress, and begin to decline under much lower stress levels, than others.

Take fine motor skills, which require the brain to send very precise instructions to the hands in order to complete movements like writing one's name or working a tumbler on a lock. This kind of task is easily disrupted by even small amounts of stress. Psychologist Alan Baddeley carried out a classic experiment on this effect in the 1960s. He had scuba divers perform a dexterity test that involved fastening screws into a plate. He found that, as the divers went deeper and they became more stressed by the cold, the isolation, and concerns about their safety, their performance got worse. They took longer to complete the task and their accuracy decreased.

It's easy to notice this effect without even getting in the water. Ever noticed how tying your shoe becomes much more difficult when you're in a hurry? Or how sluggish your tongue gets when you find yourself having to speak in front of a group? Both are examples of fine motor skills coming unglued under stress.

In contrast, gross motor skills—big-muscle actions like running, punching, jumping, and heavy lifting—are relatively resistant to stress. These involve relatively simple actions that don't need to be as delicately controlled, and they may never decline, no matter how intense the danger. When you're being chased by a lion, you probably won't have any problem getting your feet to do their stuff.

Although it's not entirely clear why these two kinds of motor actions should respond differently to stress, it's interesting to note that they are executed differently by the brain. Voluntary muscular actions—that is, those movements that we consciously choose to undertake—are initiated within a region called the primary motor cortex, which lies just beneath the crown of the skull. To carry out a gross motor movement, the motor cortex sends a signal to an intermediary neuron, which then connects to a motor neuron in the spinal cord, which finally transmits an impulse to the necessary muscle. To initiate a fine motor movement, however, the motor cortex communicates directly with the motor neurons in the spinal cord. This allows the muscular action to be controlled much more precisely. In the course of evolution, more of these direct connections became established. Humans have more than apes, and apes have more than monkeys. Among monkeys, capuchins, which have relatively abundant direct pathways, are able to pick up small objects with their nimble fingers, while squirrel monkeys, which lack such pathways, can only scoop things up with all fingers working together as one. Since gross motor skills are older, in evolutionary terms, their method of operation may be less susceptible to disruption by stress.

The difference between the fall-off in fine and gross motor skills has life-or-death consequences when it comes to armed combat. During the nineteenth century, dueling was part of the aristocratic honor code. The prescribed stance when battling with pistols was to stand very still, at a specific distance from one's opponent aim your sights at the enemy, and then carefully pull the trigger. "Be cool, collected, and firm, and think of nothing but placing the ball on the proper spot," advises an 1836 guide to the art of dueling. "When the word is given, pull the trigger carefully, and endeavour to avoid moving a muscle in the arm or hand."

The idea that one should shoot a pistol by holding it with one hand and aiming it persisted well into the twentieth century. Indeed, on the firing range, it can work quite well, especially if you brace one hand with another for stability. But research into real-world police shoot-outs reveals a surprising fact: People who fired their guns this way hardly hit anything. Instead, today's police officers are taught what's called an "isosceles shooting stance," in which the shooter faces her target squarely, holds the pistol with both hands at eye level, and stretches the arms out straight toward the target. Relying much more on large muscle groups, the isosceles is a more natural stance and far easier to execute under extreme stress.

RUNNING ON AUTO

As we saw in the last chapter, motor routines that we've done over and over again become so ingrained that even under intense stress they can unfold as automatically as a reflex. This principle, called "overlearning," has been the core underpinning of military training for as long as there has been such a thing. Flavius Josephus, a Jewish general and historian of the first century AD, ascribed the success of the Roman army to their meticulous preparation through endless practice. "Their military exercises differ not at all from the real use of their arms, but every soldier is every day exercised, and that with great diligence, as if it were in time of war," he wrote, "nor would he be mistaken that should call those their exercises unbloody battles, and their battles bloody exercises." Because the Roman soldiers overlearned every aspect of their military task set, their performance was seldom disrupted by the chaos of battle.

Poorly learned skills, on the other hand, come undone easily under the mildest pressure. As anyone who has ever stood at a wedding altar can attest, those few lines of vows that you so easily rattled off the day before can become very elusive when you're standing in front of a crowd of friends and family. That's why officiants usually prompt the bride and the groom with a few words at a time. Likewise, most of us drove much better when practicing for our first driving test than we did during the test itself, with the examiner watching our every move.

So what's the difference between the two kinds of skills? It comes down to the neurology of how we learn. Take a physical skill like hitting a tennis ball. When we're learning how to do it for the first time, we think about it consciously, using the planning circuitry that's located in our prefrontal cortex. When I'm taking my very first forehand swing, for instance, I'm consciously carrying out each step: Hold my racket like this. Plant my feet like this. Now swing my arm forward like this. Psychologists call this a "controlled" process. It's clumsy and slow—that's what being a beginner is all about.

As I play more tennis, the sequence of motions necessary to hit that forehand shot gradually becomes encoded into subconscious circuitry located in a region called the ventral striatum. Before long, I don't have to consciously think through what I'm doing: My subconscious takes care of the job all by itself, much faster and more efficiently. It can even improve on what my conscious mind has trained it to do, correcting its performance on its own through trial and error. As I play, I don't need to think about adjusting the angle of my wrist to improve my shot, or to consciously change the angle of my elbow to give the ball some top-spin. All I have to do is keep hitting balls. The more I play, the better I get. Instead of relying on a conscious, controlled process, I'm relying solely on subconscious, automatic execution.

Once a process becomes automatic, it goes on out of sight of my conscious thoughts. I can't tell *how* I hit a shot, exactly; I just do it. Some researchers have dubbed this effect "expertise-induced amnesia." The brain executes these kinds of behaviors quickly and effortlessly, even under stress.

Poorly learned motor patterns, on the other hand, haven't yet fin-ished being transferred from conscious to automatic execution. You need to plan what you're doing with the executive part of your brain and consciously monitor how the motor cortex is carrying it out. In short, you need to think about what you're doing. And as we'll soon see, under intense stress thinking about anything can be very difficult indeed.

THE ADVANTAGE OF DOING THINGS BADLY

It's a recurring theme: Under stress, the body concentrates its efforts in some areas at the expense of others. Call it the pay-the-piper effect. We've seen how cortisol shifts the way the body allocates its energy resources, flooding the bloodstream with sugar to fuel the muscles. This reallocation makes possible an incredible rise in short-term strength of the kind that Tom Boyle experienced. But it also exacts a price. The body has limited resources, and energy expended on one thing has to be borrowed from somewhere else.

For instance, the digestive system. When you're at rest, a significant portion of your blood flow travels through the network of capillaries around your intestines, absorbing nutrients from your ingested food and carrying it to the liver to be processed for storage. This is an essential function, in the big picture, but when you're facing imminent danger it suddenly loses precedence. All that blood needs to be shifted to where it's needed. In the grip of a really severe terror, the body shuts down the entire digestive system. As salivation stops, the mouth goes dry. The stomach ejects its contents through vomiting. The bladder empties, and so does the rectum. Studies conducted during World War II found that one in twenty airmen defecated during combat, as did one in five soldiers.

Sometimes the problem with the fear response is simply a question of degree. It does the right thing, but carries it too far. As stress increases past the apex of the inverted U, the physical benefits that moderate fear gives us become monstrously magnified, turning from assets to liabilities. Take the way cortisol mobilizes the body's energy resources and channels it to the muscles. Energy gets used up through the same kind of quick, repetitive muscular contractions that our body uses to generate heat when it's cold: It trembles.

Stress also cranks up the immune system, shifting it to a more defensive posture. A million years ago, danger often involved getting wounded, and a ramped-up immune system would prevent those wounds from getting infected. But as the stress increases, a side effect is to generate a conscious feeling of being unwell, just as you'd experience if you were coming down with an illness. When the body is under stress, the immune system triggers what's known as "sickness behavior." The queasiness I felt as I stood in the doorway of the skydive plane wasn't just due to my sympathetic nervous system's shifting resources away from the gut, and thereby inducing me to vomit. I was also receiving a message from my immune system, warning me that it was about to swing into survival mode.

Even a heightened sense of focus can have its downside. Moderate stress allows us to marshal all of our cognitive abilities, home in on an urgent threat, and ignore everything else. At higher levels of intensity, however, that focus tightens into single-mindedness. Psychologist James Easterbrook, who first explored this phenomenon in the 1950s, called it "perceptual narrowing." In the grip of extreme fear, we not

only ignore irrelevant sights and sounds around us; we ignore everything except the one thing we're focusing on.

This kind of tunnel vision can be a blessing or a curse, depending on your situation. It can be advantageous if you're only facing one
immediate threat, and survival depends on focusing all of your mental and physical resources on defeating it. No doubt for much of our
evolutionary past, this is the kind of situation our ancestors tended to
find themselves in: a lion closing in for the kill, an enemy warrior in
close combat. Today, we often face danger in complex, fast-changing
environments in which urgent threats might be developing on the
periphery of our awareness. If what you're focusing on is not the most
imminent threat, you're in trouble.

On February 6, 1996, a Boeing 757 took off from Puerto Plata
in the Dominican Republic heading for Germany. Soon afterward,
a blocked sensor began feeding incorrect airspeed information to the
plane's autopilot, causing it to pitch the plane's nose abnormally high.
Another automated system then reacted to this change by reducing
the power in the aircraft's engines. The flight crew, confused by the
resulting warning buzzers and the contradictory airspeed readings,
became so caught up in trying to adjust the autopilot that they over-
looked the most basic solution to the problem—to simply disengage
the autopilot and fly the plane manually. Instead, all they managed
to do was to correct the plane's nose-up attitude, while overlook-
ing the fact that the engines' power settings were still too low. The
plane lost altitude and crashed in the Atlantic Ocean, killing all 189
on board.

Intense stress causes a deterioration of cognitive processes across
the board. Earlier, we saw how noradrenaline in the prefrontal cortex
triggers the $\alpha2$-neuroadrenaline receptor, which results in increased
focus and mental energy. If stress gets too high, however, and noradrenaline rises to high enough levels, another type of receptor, the
$\alpha1$-neuroadrenaline receptor, is activated. This acts to shut down
the operation of the prefrontal cortex. Anything that requires conscious thought, deliberation, or planning becomes inordinately difficult. Laboratory studies have shown that when people are asked to

perform complex tasks under pressure, they tend to make decisions based on incomplete information and rely on oversimplified assumptions. They use rigid decision making and don't take time to consider as many options.

Crucially, they also suffer deterioration in the working memory, the short-term data store in which we hold the things that we're actively, consciously thinking about at any given moment. Normally, we can hold about seven things in working memory at once. Some people can hold more, and some less, with the capacity corresponding fairly well to a person's IQ score. In the grip of intense fear, inevitably, working memory declines. When its capacity is exceeded, solving problems and analyzing difficulties becomes nearly impossible.

To put it more bluntly, fear makes us stupid. Some call this effect "brain fuzz." It's exactly what affected the recruits in the Task FIGHTER study who botched the psychological tests as they felt the last minutes of their lives ticking away.

A TWELVE-MINUTE FLIGHT

The hazards of brain fuzz are particularly acute for pilots, who have to monitor a number of cockpit instruments, communicate with air traffic control, keep track of where they are, and on top of all that fly the airplane. One bad decision is rarely enough to kill a pilot, but it's often enough to create stress and to narrow the range of safe options. That stress can lead to more bad decisions, and an ever-narrowing margin of error that ultimately vanishes.

That's exactly what happened to Yankee pitcher Cory Lidle. The thirty-four-year-old took off from New Jersey's Teterboro Airport on an October afternoon with the aim of taking a sightseeing flight around Manhattan. It was not a great day for flying. Showers had passed through the area earlier in the day, and the weather remained gusty and overcast, with a ceiling of clouds low enough to obscure the tops of the highest buildings. But Lidle was keen to spend time in the air. An enthusiastic novice, he'd gotten his pilot's license just eight months before.

Lidle taxied onto the runway, lined up with the centerline, and gunned the 200-horsepower engine of his Cirrus Design SR-20, a four-seat airplane. Beside him was his flight instructor, twenty-six-year-old Tyler Stanger. Lidle hadn't flown in over a month, and was feeling a little rusty. Having Stanger aboard helped boost his confidence. More importantly, Stanger was familiar with the intricate airspace over New York City. As both men knew, if they wandered into the wrong patch of sky without authorization, they could be in big trouble. This hazard, added to concerns about the marginal weather, would add critically to Lidle's stress level.

As the Cirrus reached 80 mph, Lidle pulled back on the yoke and the plane floated off the runway. Lidle turned southeast toward the Hudson River, five miles away.

Even though the weather was marginal, the men were in for a spectacular scenic ride. I've flown the route myself several times, and the trick is to stay over the Hudson at an altitude of 1,100 feet or less. The Federal Aviation Administration has designated this airspace an "exclusion area." You're flying in an invisible tunnel carved out of the heavy-duty Class B airspace that surrounds the Newark, LaGuardia, and JFK airports. Class B is for the big guns—airliners flying their heavy loads of passengers at high speeds. It's no place for a small plane to be.

Heading south, the view of the city is dazzling: You're lower than the tops of the highest buildings, and as you pass over the George Washington Bridge, it almost feels like the top of it might snag your undercarriage. Soon you pass by the Statue of Liberty. If you want to, you can turn left and head up the East River. A branch of the exclusion area extends northward from Governor's Island up to the middle of Manhattan—but then it dead-ends. If you want to continue, you have to first call up LaGuardia tower and get permission to proceed into Class B airspace; over Central Park and back to the Hudson River.

Lidle and Stanger chose a simpler plan. Flying at low altitude—about five hundred to seven hundred feet—they banked around the Statue of Liberty and cut east toward the mouth of the East River without calling LaGuardia. Without permission to continue on into Class B airspace, they were committed to coming back out of the East River exclusion the way they'd gone in.

In my experience, the most striking thing about flying up the East River is just how narrow it is—in places, less than two thousand feet. And it's winding, with a dogleg to the right at the Brooklyn Bridge and then another to the left at the Brooklyn Navy Yard. It would have seemed even narrower to Lidle, flying at just five hundred feet, with a low overcast above him and the ranks of skyscrapers lining the Manhattan shoreline like the walls of a man-made canyon. As a novice pilot, he likely felt intimidated. And on top of that Lidle was very soon going to have to make a challenging steep turn to reverse course. There wasn't much time to plan ahead: traveling at 112 mph, the Cirrus would cover the length of the exclusion in less than four minutes.

Time pressure is a form of stress that can be every bit as incapac-itating as fear. When we have too much to do, and too little time to do it in, it can feel like our thought processes have frozen up. Pilot Dick Rutan, the first man to fly around the world nonstop, says that when he's in that kind of situation "my mental task box is saturated." I've been in that position myself in the cockpit, and it's a baffling sen-sation that brings on a state of paralysis. It simply doesn't occur to you to do things that, in retrospect, seem utterly simple and obvious. Unfortunately for Lidle, by the time he was flying up the East River, his margin of error was tissue thin. A single mistake could kill. As it happened, he made two.

At the speed Lidle was flying, the Cirrus requires at least two thousand feet of latitude to complete a full 180-degree turn. The river at that point is only 2,100 feet wide. The turn was physically possible, if done exactly right. But here Lidle made his first critical error. In order to give himself maximum clearance from the obstacles on either side, he had instinctively drifted to the center of the channel. When he began his turn, he had already given up nearly half of his available space.

At this point, Lidle's situation had come down to one very simple question: Which way to turn? Left or right? And here's where the pitilessness of brain fuzz comes to the fore. With the comfort of hindsight, it's absolutely obvious what the correct answer is. On the Manhattan side of the river, the buildings are tall; higher indeed than the altitude that the Cirrus was flying at. On the Brooklyn side, they're lower. So Lidle should have turned right. If he failed to make the turn, all he'd bust would be airspace. Sure, he'd have some explaining to do to the FAA. But airspace is a lot softer than buildings.

There's another factor that would have reinforced that decision, if Lidle had the time to think about it. The storm that had swept through that morning had brought with it a shift in the wind. Instead of blowing from the southwest, as it usually does, the wind was now coming from the northeast. As Lidle made his turn, the wind would be blowing him toward Manhattan. If he turned to the right, it would tend to keep him over the river. If he turned to the left, it would blow him further toward the buildings on the Manhattan shoreline.

Lidle didn't have time to take these factors under consideration. When we're under intense stress, we fall back on certain rules of thumb, guidelines that help simplify the options we're facing. These rules help us make quick decisions in situations where indecision can be fatal. And the rule that Lidle fell back on is a basic one familiar to all pilots: Given a choice between left and right, the natural choice is always to turn left.

Why? Simply because the pilot's seat is on the left side of the aircraft. When you turn left, you can look out the window by your shoulder and see where you're going. When you turn right, your view is largely obstructed by the bulk of the aircraft. Instinctively, we want to see where we're going. Especially when it's full of solid things we could potentially run into.

So Corey Lidle turned left. He had nearly completed the turn when he ran into the north side of a fifty-story apartment building at an altitude of 333 feet. He and Stanger were killed instantly. Each left behind a widow and a small child. From time of takeoff, the entire flight had taken twelve minutes.

OVERWHELMED

Poor decision making is a potential hazard even when we're under moderate stress. Raise the stakes still higher and the mind can melt down. Extreme arousal combined with cognitive narrowing equals a kind of hysterical fugue state, as the hyper-activated amygdala shuts down the cortex and takes control.

Among hunters, this frame of mind is known as "buck fever." When a novice shooter has quarry in his sights for the first time, the mix of powerful emotions—excitement, bloodlust, fear—can be so overwhelming that strange behavior results. Around the campfire, you'll hear stories of first-time hunters shouting "Bang! Bang!" at their prey, or running after it trying to club it with their gun, or trying to run to tag a felled deer without remembering to first climb down from their tree stand. More commonly, novice hunters simply find them-selves firing wildly, or forgetting to fire at all. During his first attempt to bag a deer, hunting guide Russ Chastain writes that he was reduced to "a wheezing, trembling, adrenaline-filled idiot with knees of Jell-O and one hell of a heart rate—until the deer calmly walked away."

Ordinary citizens who suddenly find themselves in the midst of a life-or-death crisis can be particularly vulnerable to this kind of cogni-tive shutdown. Unprepared for the overwhelming force of their brain's full-blown fear response, they can be reduced to abject hysteria, or locked in a kind of catatonic trance. Before they can help them, rescue personnel might have to break through their psychic shell by shout-ing obscenities at them. Airline flight attendants are now trained dur-ing emergencies to scream at passengers who aren't moving quickly enough toward the exits in order to knock them out of their fugue state and get them active again.

Amidst the clamor of battle, it can be impossible to get terrified soldiers to undertake any course of action they haven't trained for. Those lost in a fog of fear can become stuck in an endless loop. Unable to summon the mental resources to take stock of the situation and plan an efficacious course of action, they repeat a single well-learned skill over and over, despite its manifest futility. Called "perseveration," such behavior is also seen in patients with damage to the prefrontal cortex of their brain. Of the twenty-seven thousand muskets recovered from the field after the Civil War battle of Gettysburg in 1863, 90 percent were found to be loaded. Of these, nearly half had been loaded more than once. Evidently the soldiers who carried them had gone through the laborious process of loading them again and again without ever managing to fire. Amid the deafening chaos of battle, the men simply resorted to what they had been trained to do. One gun was found to have been loaded twenty-three times.

SHUTTING DOWN

When fear escalates beyond what is bearable, it's possible to simply remove oneself psychologically from the scene.

Johan Otter's ordeal began with an unexpected flash of fur. Before he knew what was going on, Otter was pushing forward on the narrow trail, stepping in front of his daughter to shield her from the danger he only blurrily perceived. *Brown fur. Ears back. Teeth bared.* A quarter second, and the thing had crashed into him, its teeth sinking into the flesh of his thigh. He looked down. There was a grizzly bear clamped onto his leg. It didn't seem real. He thought: *So this is what it feels like to have your flesh ripped.*

Otter, forty-three, had come to Glacier National Park with his teenaged daughter, Jenna, on a relaxing vacation in late August. They'd risen early, and by 7:30 a.m. were midway up a steep trail that dropped off to one side in a precipice several hundred feet high. Far below, Grinnell Lake glistened in the cool autumnal sun.

Then they'd turned a corner and found themselves five feet from a mother grizzly bear. Jenna, who was walking in front, turned back, screaming, and ran past him. Otter stepped between the bear and his daughter and tried to beat the animal off. It bit him over and over again, snapping at his legs and arms, grabbing him and shaking him. He tried to curl into a fetal position, but the bear wouldn't let him. He spotted a thimbleberry bush jutting from the slope below him. Pushing away from the grizzly, he rolled to the side and plunged toward the bush. He landed safely, but the bear came barreling after him. In seconds it was on him, jumping on his back, ripping at his day pack.

"Time was really weird at that point," Otter says. "It was moving really, really slow." His predicament seemed so unlikely, so unreal, so unexpected, that even as the bear began sinking its teeth into his scalp he felt as though he were watching himself perform in some kind of a movie. "I was even thinking at one point, 'Okay, they can stop the cameras right now, and we're done with the stunt, because it's really becoming annoying, they can pull the bear away from me.'" Otter says. "But of course nothing was stopping, because this was real. It was like a dream but it wasn't a dream."

What Otter experienced that day was a kind of narrowing so intense that it was as though he had removed himself from the picture entirely. Called dissociation, this sense of watching oneself from a distance, or of being in a movie, is a short-term coping response. Lost in a sort of trance, a person is saved from having to deal with the emotion that might otherwise engulf them.

For Otter, the sense of being removed from his own body didn't prevent him from taking steps to save himself as he felt the bear's massive jaws lock on his head. "I felt a tooth going into my skull and I thought, 'This is going to be it.'" Otter says. There was only option: to break free of the bear and jump off the cliff. It was several hundred feet to the bottom, but to Otter the choice was clear. "It sounds very weird, but I was very methodical," says. "There was no room for emotions. It was like, 'This is what's a better option now.'"

He pulled away and plummeted down the cliff. By chance, a small ledge protruded from the side of the mountain twenty-five feet down, and it stopped his fall. The bear peered down at him, unable to follow. Eventually it wandered off.

Otter was in bad shape. His scalp was ripped off, his arms were torn down to the tendons, his neck was broken in two places, and he was losing blood. A stream of cold rainwater runoff drenched him. Worse, he didn't know how badly Jenna had been hurt. Yet he felt oddly elated. "It was much more important that I was alive. You kind of have this euphoria, in a way, that you're still alive."

After an hour, passing hikers found the pair and called for help. Six hours after the attack, Johan and Jenna Otter were medevaced off the mountain by helicopter. A long road of recovery and rehabilitation lay ahead for Otter, including three months with his head wired in a metal halo.

Throughout the whole ordeal, Otter says, he wasn't conscious of any feelings of fear. "There wasn't any room for fear. I had to intellectualize myself, basically, through that whole event. I was not afraid until three months later, when my halo was taken off. Suddenly I didn't have that external protection around me anymore, and I was like, okay, I am so afraid right now. I remember saying [that] to my wife, and I just started crying, 'I am so afraid. I should have been so afraid.' I remember saying that: 'Man, I should have been so afraid, and I totally was not.'"

As he looks back on the experience, Johann Otter has a hard time making sense of how and why his brain worked the way it did. Many survivors of life-and-death dramas come away equally baffled. The strange effects of fear on the mind can seem like a chaotic whirlwind when we're experiencing them. In extremis, our fear centers take over totally, and our conscious mind loses control. We find ourselves behaving in ways that we cannot predict or understand. But in reality, the madness of fear does have a method all its own. There is a hidden order that underlies the seeming chaos of fear.

PART TWO

THE STRUGGLE FOR CONTROL

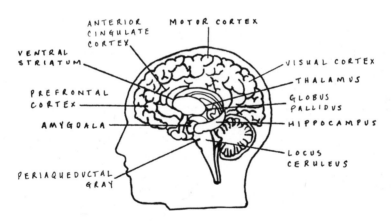

The fear response unfolds largely out of sight of the conscious mind, in ancient, deep-lying structures such as the periaqueductal gray. It is here that the stereotyped behavioral responses to intense danger—freezing, fleeing, fainting, and fighting—are generated.

Art by Sandra Garcia

CHAPTER FOUR

THE STRUCTURE OF CHAOS

IN THE THROES of intense fear, we suddenly find ourselves operating in a different and unexpected way. The psychological tools that we normally use to navigate the world—reasoning and planning before we act—get progressively shut down. Instead, in the grip of the brain's subconscious fear centers, we behave in ways that to our rational minds seem nonsensical or worse. We might respond automatically, with pre-programmed motor routines, or simply melt down. We lose control.

In this unfamiliar realm, it can seem like we're in the grip of utter chaos. But although the preconscious fear centers of the brain are not capable of deliberation and reason, they do have their own logic, a simplified set of responses keyed to the nature of the threat at hand. There is a structure to panic.

The first researcher to begin sketching out the logic of fear was Harvard physiologist Walter Cannon. In 1915, he pointed out that the different effects of sympathetic arousal—the increased heart rate and blood flow, the sweating, the trembling, and all the rest—all serve one underlying purpose: to prepare the body for a vigorous defense. Cannon's idea was so persuasive that his pithy encapsulation—"fight or flight"—has become the best-known term for the sympathetic nervous system.

Like many theories that subsume a great deal of data into one compelling explanation, however, it has turned out to be oversimplified. There are not two kinds of defensive reaction, but rather at least four, each with a suite of physiological responses optimized to handle a different category of threat. When the danger is far away, or at least not immediately imminent, the instinct is to freeze. When danger is approaching, the impulse is to run away. When escape is impossible, the response is to fight back. And when struggling is futile, the instinct is to become immobilized in the grip of fright. Although it doesn't slide quite as smoothly off the tongue, a more accurate description than "fight or flight" would be "fight, freeze, flight, or fright"—or, for short, "the Four F's."

On a winter morning a few years back, a young woman named Sue Yellowtail went through them all in about ten minutes.

FREEZE

The Mancos River rises in southwestern Colorado and flows through the Ute Mountains on its way to New Mexico, where it empties into the San Juan River three miles shy of the Four Corners intersection. Over millions of years, the river and its tributaries have carved a fan-like rill of dramatic canyons out of the ancient sediments of the Mesa Verde tablelands, a maze of vertiginous stone walls. The rugged, arid landscape of juniper forest proves a rich habitat for wildlife.

At twenty-five, Sue Yellowtail was just a few years out of college, working for the Ute Indian tribe as a water quality specialist. Her job was to travel through remote areas of the reservation, collecting samples from the streams, creeks, and rivers. She spent her days crisscrossing remote backcountry, territory closed to visitors and rarely traveled even by locals. It's the kind of place where, if you got in trouble, you were on your own.

On a clear, cold morning in late December, Yellowtail pulled her pickup over to the side of the little-traveled dirt double-track, a few yards from a simple truss bridge that spanned the creek. As she collected her gear she heard a high-pitched scream. *Probably a coyote killing a rabbit,* she thought. She clambered down a steep embankment to the water's edge. Wading to the far side of the creek, she stooped to stretch her tape measure the width of the flow. Just then she heard a rustling and looked up. At the top of the bank, not thirty feet away, stood a mountain lion. Tawny against the brown leaves of the riverbank brush, the animal was almost perfectly camouflaged. It stared down at her, motionless.

She stood stock still.

Yellowtail had entered the first instinctual fear-response state, the condition of freezing, otherwise known as "attentive immobility." Even before she had become aware of danger, subconscious regions of her brain were assessing the threat. Cued to the presence of a novel stimulus, the brain deployed the orienting reflex, a cousin of the startle reflex. Within milliseconds, Yellowtail's heart rate and breathing slowed. A brain region called the superior colliculus turned her head and skewed her eyes so that the densest part of the retina, the fovea, formed a detailed image of the cat. The visual information then flowed via the thalamus to the visual cortex and the amygdala. Her pattern-recognition system found a match in the flow of sensory information. It recognized a pair of eyes. Then the outline of a feline head. In less than half a second, before her cortex even had time to complete the match and recognize what she was seeing, her emotional circuitry had already assessed the situation: It was bad. Subconsciously, her brain also determined that the threat was not immediately pressing, and so a region called the ventral column of the periaqueductal gray (vPAG) triggered attentive immobility. This is generally considered the first stage of the fear response, because it tends to occur when the threat is far away or not yet aware of the subject's presence. The goal is to keep it that way.

When a person is frozen with fear, she is motionless, but far from passive. Coursing with cortisol and adrenaline, the body is primed for physical action, the mind alert and intensely focused. The heart rate slows and blood pressure shoots up. Muscles tense, and the pupils dilate. The body may tremble and the eyes bulge. If the fear is intense, the mind might be plunged into a state called hypervigilance, in which a person scans the environment rapidly and randomly, unable to think clearly through the available options.

Many prey animals have particularly well-developed freeze responses. Deer, famously, will stand in the headlights of an oncoming vehicle and watch motionless as they are about to be run over.

Freezing is a posture of an animal that, while in danger, is primarily concerned with not getting into worse danger. Its plan is to do nothing, hope to avoid being detected, and see what happens. In the natural environment, it often proves an effective strategy. Young antelopes can spend the better part of the day lying crouched and motionless in tall grass, their ears tucked and head pressed against the ground. When accidentally disturbed by a passing lion or hyena, they bolt so unexpectedly that the predator may be too startled to chase after it.

The freeze response, however, doesn't engage because it's a rational reaction. The brain's subconscious fear centers activate it. And as we've seen, since these automatic processes rely on simplified logic, sometimes, in the context of modern living, they respond inappropriately. Say you're about to take a recreational skydive. Your frontal cortex looks down through the open door in the fuselage and tells itself that, based on well-known statistical analysis, the actual threat to your well-being is minimal. Your amygdala, however, looks out the doorway and sees the situation through a more primitive lens: Great height! Danger! Freeze! And so you wind up clinging for dear life to the edge of the doorway while your instructor tries to pry your fingers away.

One frequently hears stories about inexperienced climbers freezing halfway up a rock wall, or performers going rigid in terror on stage. In most such circumstances, the worst that results is some embarrassment. But in other situations the freeze response can kill.

On August 17, 2006, Dr. Greg Keating, a sixty-three-year-old biology professor and dean at SUNY Upstate Medical School, took off with a flight instructor in his Scheibe motorglider from Canandaigua Airport in upstate New York. After flying some basic maneuvers, the men returned to the airfield to practice takeoffs and landings. On the third takeoff, the examiner waited until the plane had climbed to four hundred feet above the runway, then cut the power to simulate an engine failure.

Keating didn't expect this, and, although he'd practiced such maneuvers in the past, no doubt he felt a surge of tension. I've performed similar exercises myself, and the sudden loss of power always provides a flash of fear. Time is short; altitude is limited; the necessary procedure must be carried out quickly and correctly.

The first thing to do in such an emergency is to lower the plane's nose to gain airspeed. If the pilot hesitates and loses too much speed, the plane will plummet uncontrollably. As it happened, Keating started the turn, but he failed to lower the nose sufficiently. His airspeed began to drop. Sensing that the wing was about to lose lift, the examiner shouted, "I've got the controls!"

Keating's hands stayed gripped around the stick. He had frozen in fear. The examiner fought to push the stick forward to lower the nose, but it was too late. The rush of air went quiet as the plane slowed dangerously, trembled, and lurched forward, falling toward the farmland below like a roller coaster at the top of its first hill. Seconds later, the plane crashed into a tree and tumbled into a nearby field. The impact crushed the nose to the back of the front seat. Keating was killed. The FAA examiner survived with minor injuries.

As it happened, Sue Yellowtail did not need to overcome instinct. Hers was just the kind of situation that the behavior had evolved for: eluding a nearby predator. Freezing was an entirely appropriate tactic. But freezing is essentially a temporary measure, a stopgap until the danger either goes away or becomes more pressing. It's a posture that asks the question: What next?

In the morning light of the Mancos Canyon, human and animal stood confronting each other. Yellowtail had never seen a mountain lion in the wild before. Even as she fought to contain her fear, she marveled at the beauty of it. Its dark eyes looked back at her. Who knew what it was thinking, behind that gaze? Was it curious, or hungry?

FLIGHT

As Yellowtail locked eyes with it, the mountain lion moved forward, descending the shrubby bank and heading straight toward her.

Being spotted by a predator, or approached too closely, is generally sufficient to break the spell of the freeze response. In Yellowtail's case, the moment she realized that the big cat was heading for her she began moving away, wading back across the three-foot-deep stream toward her truck. For prudence's sake, she thought, she'd better keep the width of the icy stream between her and the animal. As she made it to the far side, the mountain lion quietly slipped into the water.

A former biology major, Yellowtail had studied predator behavior. She knew that if she began climbing the steep bank up toward her truck then she would expose her back, and she guessed that the moment of vulnerability might spur the mountain lion to attack. Instead, she moved quickly down the edge of the stream and crossed again, feeling her way over the slick cobbles underfoot. Looking behind her, she expected to see the animal climb the far bank and disappear. But no: It followed her path along the water's edge and again started swimming after her.

I'm in trouble, Yellowtail thought. *This is serious.* There was no doubting the mountain lion's intention now. Trapped between the stream's steep and narrow banks, she couldn't think of any way to keep the animal away. She was holding a microcassette recorder that she kept for taking notes, and she threw it at the cat. It just kept coming.

Yellowtail retreated down the riverbank, shouting and throwing rocks and chunks of ice. Somehow she managed to keep herself from running. She crossed the stream, walked further down the bank, and crossed again. The cat followed, relentlessly closing the distance. Even as she felt a sense of panic building, Yellowtail had enough presence of mind to understand that what she was seeing was a classic example of predator behavior. Running would only stoke the animal's attack instinct. She had to fight the urge.

The mountain lion was close now, near enough to pounce. As she splashed once more across the stream, the need to run surged over her like a shiver. She bolted, splashing madly through the shallow water, her legs churning over the rough, slippery cobbles of the stream bed.

She ran with everything she had.

Yellowtail was now in the grip of the second phase of the fear response: flight. The sudden movement of the mountain lion had broken the spell of her attentive immobility and got her moving. While the animal was still a fair distance off she had managed to keep her wits and suppress her fear centers' automatic panic reaction. She was able to assess her situation, to weigh options, to make a plan and to carry it out, despite the urgings of her rising fear. But, as the cat drew closer, the balance of power between her amygdala and her frontal cortex shifted. Reason and willpower wavered as the fear grew stronger. At last they gave way altogether.

This process has been witnessed in the laboratory using brain-scan technology. Subjects inside a functional magnetic resonance imaging (fMRI) scanner were asked to play a game that resembled Pac-Man, in which they were chased by predator. When they were "caught" they were given a series of mild electric shocks. While not exactly a realistic scenario, the game did elicit brain activity that paralleled Yellowtail's. When the "predator" was far away, the subjects' brains showed activity mostly in the prefrontal cortex. As it drew nearer, the area of greatest metabolism shifted to the periaqueductal gray, the region that codes for the behavioral patterns of the Four F's.

As we saw with freezing, what works well out in the wild can be disastrous in the modern world. The urge to run can be deadly. In some of the past century's most catastrophic fires, the leading cause of death was not fire, or even smoke inhalation; rather, it was being crushed to death in a panicked stampede. There doesn't even need to be an actual threat, so long as the crowd believes that there is one. Among the worst mass fatalities during the German bombing campaign against London in World War II was an incident that took place when a crowd of civilians lined up outside the shelter in the Bethnal Green Underground Station. An air-raid siren had sounded several minutes before, but the crowd didn't panic until a nearby anti-aircraft battery fired a salvo of rockets. A woman carrying an infant stumbled at the top of the steps leading down into the station; the surge of the crowd behind them caused hundreds more to fall. Amid the din of the guns and rockets, no one could hear the screams of the men, women, and children being crushed to death. Those at the back, unable to move forward and fearing that they were being forcibly excluded from safety, pushed harder. Within fifteen minutes, 173 people had died. In a twist of grim irony, not a single bomb had fallen.

Unlike civilians, military personnel are trained to overcome the urge to flee danger. This was the point of the Roman army's ceaseless training: To replace the flight instinct with overtrained motor routines that would serve them well in combat. Among a less disciplined army, panic could spread like a virus on a battlefield, leading to disaster. In the majority of battles throughout history, victory went to the side not that killed the most enemy, but to the one that withstood longest the urge to give up and run. "Valor is superior to numbers," wrote the Roman historian Vegetius in 390 AD. "A handful of men, inured to war, proceed to certain victory, while on the contrary numerous armies of raw and undisciplined troops are but multitudes of men dragged to slaughter."

The Union Army learned this lesson during the first major engagement of the Civil War. It was the summer of 1861, and after three months of hostilities North and South alike were still scrambling to put themselves on a war footing. Each fielded armies that as yet were small and ill-trained; each still hoped to win quickly and with relatively little bloodshed.

In late July the Union commander, Brigadier General Irvin McDowell, marched a force of thirty-five thousand men against the Confederate army camped near Manassas, Virginia. Crossing Bull Run Creek, the Yankees drove back the defenders from a hilltop position but failed to take advantage of the enemy's disorderly retreat. The Confederates were able to reestablish a defensive line atop the next hill. Reinforced with newly arrived troops and artillery, the Confederates managed to overrun an artillery battery at the southern edge of the opposing line, and the Union forces began to fall back. Their retreat was orderly at first, but when Confederate cannon blew up a wagon that was trying to cross back over Bull Run, nearby troops panicked. The contagion of fear was unleashed. Retreat turned into headlong flight as soldiers ran for their lives, throwing away guns and equipment.

To make matters worse, the road to Washington was jammed with spectators who had come out to see what they assumed would be a great Union victory. Chaos ensued. The fleeing artillery lost all but one of their cannon. Officers tried in vain to rally their troops, who did not stop running until they got to Washington. More than a thousand Union troops were captured in the debacle, with more killed or wounded. For the North, what was worse than the casualties was the awful realization that the war would not be won quickly, or without great bloodshed. Both sides settled in for a long, hard fight.

Yellowtail's desperate attempt to flee from danger was little more successful than the Union Army's. She only made it halfway across the creek before her rubber boot caught on a large rock. She stumbled, twisting, and went down hard into the water. At that instant the mountain lion pounced. Instinctively, it lunged for Yellowtail's neck, but as she fell it misjudged and sank its teeth into her scalp. Under the weight of the big cat, Yellowtail slipped below the surface.

FRIGHT

Looking back on that moment, years after the fact, Yellowtail can still recall every detail with perfect clarity. She remembers feeling the warmth of the animal's mouth on her head. She remembers looking up toward the surface through her sunglasses and thinking, with a perplexing degree of calm: *When your time's up, your time's up.*

Yellowtail had entered a third modality of the fear response, a state known as "tonic immobility," or quiescence—in lay terms, "playing possum." When an animal is seized by an attacker, the caudal ventrolateral region of the vPAG generates a response that, from the outside, looks like total collapse. In the teeth of a full-blown sympathetic response, the parasympathetic system now swings into overdrive. The body, insensitive to pain, goes completely limp, often falling to the ground with limbs splayed and neck thrown back. Eyes closed, it trembles, defecates, and lies still. It looks, in a word, dead.

The hope of quiescence is that the attacker, thinking its quarry has expired, will stop attacking. Indeed, many predators will not eat prey that looks dead; hawks, for instance, will starve to death if unable to attack moving prey. The famed nineteenth-century missionary David Livingstone was a beneficiary of this effect when he was charged by a lion he'd shot at during a hunting trip in Africa. The animal grabbed him in its jaws and shook him like a rag doll. To his surprise, Livingstone found that he felt no pain, and that indeed it caused "a kind of dreaminess." Fortunately for him, the immobility response worked as intended, and the lion dropped him in favor of some other hunters who were moving nearby.

Quiescence is the most paradoxical form of terror. With both branches of the autonomic nervous system at full throttle, the organism is both utterly relaxed and completely alert and ready for action. Pupils are dilated, breathing and heart rate rapid. Though paralyzed, incapable of voluntarily action, the animal can suddenly spring to life and flee if the opportunity arises. If quiescence goes on too long, however, heart rate and blood pressure can plunge dramatically, indeed to the point of death. It's not just folklore: You really can die of fright.

Walter Cannon proposed that this phenomenon might explain the demise of indigenous tribesmen who believe themselves cursed by witchcraft. In a 1942 paper entitled "'Voodoo' Death," he relates several such incidents, including an account by adventurer Arthur Glyn Leonard of a trip to the Lower Niger: "I have seen Kru-men and others die in spite of every effort that was made to save them, simply because they had made up their minds, not (as we thought at the time) to die, but that being in the clutch of malignant demons they were bound to die." Cannon surmised that intense fear can cause such a catastrophic drop in blood pressure that the belief in one's death can became self-fulfilling.

Virtually all animals exhibit a form of quiescence as part of their natural repertoire of behavior. Turkeys will go immobile if their heads are tucked under their wings. Laboratory rabbits are quieted for handling by holding them upside-down in a v-shaped trough. If you ever find yourself wading through a Florida swamp, it might be useful to know that alligator wrestlers use quiescence to make their opponents more tractable. When I was growing up, my brother and sisters and I would entertain ourselves before a summertime lobster dinner by balancing the crustaceans on their heads and stroking their folded-over tails until they became still. We called this "hypnotizing" them, and imagined it would make immersion in boiling water less painful. (In retrospect this seems unlikely; while quiescence does numb pain, the shock of the boiling water seemed to knock them out of it, and as they cooked they would thrash and rattle the lid.)

Quiescence is the ultimate expression of desperation, and accounts by human beings inevitably involve the ugliest kinds of trauma. Over fifty percent of rape victims report being paralyzed during the attack. Survivors of mass shootings also describe playing dead.

On the morning of Monday, April 16, 2007, Virginia Tech student Seung-Hui Cho entered the second floor of a classroom building and walked from classroom to classroom, shooting everyone he saw. Within minutes, thirty-three people would be dead. Junior Clay Violand was in an intermediate French class when he heard what he thought was the sound of loud banging. His teacher, Jocelyne Couture-Nowak, looked outside to see what was going on, then quickly retreated. Hastily, she tried to push her desk against the door to form a barricade, but it was no use. Cho pushed through the door and shot her dead.

Violand threw himself underneath his desk as Cho shot one student after another, taking his time to aim and to reload between clips. Victims moaned and yelled around him. Recounting his experience later, Violand wrote that "shot after shot went off and I never felt anything. I played dead and tried to look as lifeless as possible." Violand wondered what it was going to be like to die. He wondered if he had already been shot, since his whole body felt numb. He tried to move and found that with effort he could wiggle. When the gunman left the room, one of the other students said that help would be arriving soon. Violand whispered, "Play dead. If he thinks you're dead then he won't kill you." This proved wishful thinking. Soon Cho returned and proceeded to shoot the wounded again and again. Violand lay still once more, and his luck held. By the time he was done, Cho had shot everyone in the classroom except Violand, wounding four and killing twelve.

It's interesting to note that Violand described playing dead as a choice he had consciously made. He had no idea that he had been hijacked by his PAG. Finding himself lying under his desk, he naturally intuited that it was because he *wanted* to lie under the desk.

In Yellowtail's case, the mountain lion appeared to react to her quiescence in the same way that Livingstone's lion did to his. Momentarily, it released its grip. That was enough. In an instant she had snapped out of her dissociative dream state and was sputtering back up to the air. Without reason, without thought, she started running, flailing so hard that she ran right out of one of her hip boots.

And then—nothing. Whatever happened next, Yellowtail has no idea, because for the next ten or fifteen seconds she was overcome by a panic so blind that she blacked out. She had entered a realm of fear strong enough to shut down the memory-forming hippocampus, and perhaps even consciousness itself.

The science behind that kind of amnesia remains murky, because such intense fear is a state as yet inaccessible to science, being far beyond what researchers like Mujica-Parodi or even Task FIGHTER could evoke from their test subjects. It is known that amnesia often accompanies extremely terrifying experiences. People who experience tonic immobility, for instance, frequently have gaps in their memory about what happened. Chances are, an overdose of cortisol or a related substance, corticosterone, disrupts the hippocampus and inhibits the formation of new memories. This could be beneficial if it prevents later traumatic recollections. In fact, researchers in Israel have found that giving shots of corticosterone to mice after they've been exposed to trauma reduces their likelihood of later suffering a stress response.

Yellowtail will never know what terror her amnesia cloaked. At any rate, it did not last long—just fifteen to twenty seconds, she estimates. The next thing she remembers, she was on the riverbank on the far side of the stream. She had emerged from her blind panic oddly collected, and remembers that time seemed to be moving in slow motion. She found herself lying on top of the mountain lion's shoulders, her right arm thrust down its throat. She looked down and saw that the animal's jaws were so huge that its canines overlapped on either side of her arm.

FIGHT

I've got to kill this animal or it's going to kill me, she thought. She happened to be wearing a fly fishing vest from which hung a surgical steel hemostat on a retractable string. In the strange, lucid clarity of total fear, she reasoned through a course of action. First she tried to wrap the string around the cat's throat to strangle it, but abandoned that plan when the cat thrashed, slashing its teeth dangerously close to her fingers. Before moving on to a new strategy, she paused and carefully inspected her left hand to make sure her fingers were all there. "Because if they weren't, I was going to pick them up and put them in my pocket," she says today. "It's just crazy the stuff that you think about."

Her next thought was to stab the cat in the eye with the hemostat. "It just dawned on me: I've got to get to the brain, so the eye was the best bet." Without thinking twice, she clutched the hemostat and stabbed it over and over again into the cat's left eye. The beast screamed a horrifying yowl. She kept stabbing.

Yellowtail had worked her way through to the fight, or aggressive defense, response. Like quiescence, aggressive defense is a tactic of last resort. Often, it takes place after an animal has been grabbed by an attacker. But not always: Mothers can be extremely vicious in defense of their young, as Jack Otter found when he came between a grizzly and her cub.

Like the Beserkers we met in Chapter Two, people in the throes of full sympathetic overdrive are capable of totally uninhibited, blind violence. They will use any weapon and inflict any injury they can. On the battlefield, this impulse may be useful in the heat of fighting, but it can also lead to reckless, even mindless, behavior. And it's very difficult to shut off. Once the cortex has yielded control to the PAG, there's no getting it back until the shouting is over. The annals of military history are filled with tales of soldiers who kept slaughtering well after the battle was over. The Old Testament book of Joshua describes how the Hebrews dealt with the defeated city of Jericho: "And they utterly destroyed all that was in the city, both man and woman, young and old, and ox, and sheep, and ass, with the edge of the sword."

The technology of modern warfare has changed, but the ancient fear centers have not. Military commanders must walk a narrow line between encouraging aggression and endorsing war crimes. "When we meet the enemy, we will kill him. We will show no mercy," General George Patton told his troops on the eve of battle in Italy in 1943. "If you company officers in leading your men against the enemy find him shooting at you and, when you get within two hundred yards of him, and he wishes to surrender, oh no! That bastard will die! You must kill him. Stick him between the third and fourth ribs."

In a civilian setting, blind violence driven by fear is even more problematic. Police officers have to operate in a high-cortisol, high-adrenaline environment day in and day out without letting their Four F's getting the better of them. They might conduct an extremely dangerous high-speed car chase, followed by an on-foot pursuit, followed by a hand-to-hand struggle, without getting overwhelmed by the defensive combat response. "You can fight, but not like a demon, unless your opponent is fighting like a demon as well," writes police psychologist Alexis Artwohl in his book *Deadly Force Encounters*. "While you can use force, even deadly force, you can use only enough to stop the threat. This is not an easy task when someone has just tried to kill you and you think that this might be the end."

Add in additional stressors such as darkness, situational ambiguity, fatigue, and the perceptual and cognitive distortions caused by the threat on the ground, and it's hardly surprising that police officers sometimes use excessive force. Sometimes, wildly excessive.

In November 2006, a team of undercover New York City police officers were staking out a bar in Queens called Club Kalua, a suspected front for drugs, prostitution, and illegal guns. An officer heard a customer get into a dispute with a stripper. One of the man's friends threatened to "get my gun." They proceeded to leave the club with six other men. Fearing that mayhem was about to break out, the plainclothes officers followed the men outside and saw them get into two cars. One was being driven by twenty-three-year-old Sean Bell. The police later testified that when they asked him to raise his hands, Bell instead drove at them, hitting one officer, then backed up and rammed a police van. The police opened fire, unleashing a salvo of 50 shots that killed Bell and badly wounded two of his companions. One of the officers had fired 11 shots; another had emptied his clip, reloaded, and emptied it again, for a total of 31 shots.

Protesters against police brutality thundered through the city for weeks. Social activists claimed that this kind of overkill was a recurring behavior by the New York City Police Department. Again and again, teams of officers had gotten caught up in seemingly reflexive spasms of "contagious shooting": Under the stress and uncertainty of a potentially lethal encounter, a single police officer's initial gunshot can trigger aggressive firing by her companions, as each assumes that the others must have a good reason to be shooting. This spontaneous groupthink immediately results in a self-reinforcing spiral. In 2005, police had fired 43 rounds at man in Queens. In July 2006, three officers fired 26 rounds at a vicious dog that had bitten one of their colleagues. And, of course, famously, there was the Amadou Diallo case, in which four plainclothes officers fired 41 rounds at the West African immigrant standing in a doorway in the Bronx, after they saw him reach for what they thought was a gun. It turned out to be a wallet.

In the wake of protests, the police commissioner asked the RAND Corporation to investigate the department's problem with contagious shooting. The following year, the RAND team issued a 114-page report that recommended, among other things, that the department provide firearm training specifically aimed at preventing contagious shooting. For instance, it proposed that officers undergoing training at the shooting range should hear shouts of "He's got a gun!" and the sounds of other guns going off, in order to train them that hearing such things is not in itself reason enough to open fire.

In Yellowtail's case, there was no need to restrain her impulse to violence. She attacked the mountain lion with all that she had, until she sensed that it had had enough. She kicked off her other hip boot and got ready to stand up. The cat let go of her arm. As soon as Yellowtail's weight was off it the cat stood up, too. Yellowtail lunged at it, swearing and shouting: "Come on, you want some more?" The cat didn't move. Yellowtail lunged again, to see if it seemed ready to attack her again.

It just stood there, looking dazed. Yellowtail backed up about twenty feet, then turned and ran down the bank until she found a cattle path through the brush back up the embankment to the road. That led "The whole time I was worried that it was going to come through the brush and get me," Yellowtail says, "but it never did."

Yellowtail got into her truck and drove for help. Not until she was in an ambulance did her multiple cuts and bruises begin to throb with pain. Trackers returned to the site of the attack, located the mountain lion, and shot it. It turned out to be an aged female, underweight and weak from starvation. Yellowtail figures that if the cat had been a full-sized male she'd be dead now. As it was, she came very close.

"It's amazing what you can do when you're under that kind of stress, in a life-or-death situation," she says. "You do whatever it takes to keep yourself alive."

CHAPTER FIVE

FEAR ITSELF

OUR ANCESTORS EVOLVED their complex fear-response circuitry over hundreds of millions of years to help them survive myriad lethal threats. It was an essential piece of machinery; without it, they never would have stuck around long enough to procreate. But the fear response is not an unalloyed blessing. Once the alarm system goes into action, it can become a threat in its own right, commandeering the resources of the brain and body in ways that are beyond conscious control. You're loaded for bear whether you like it or not. And so when you're in serious danger, you've got two big things to worry about: the threat, and your own fear response.

In many cases, the fear can be worse than the threat. There are even times when a full-blown fear response erupts in the absence of any danger at all. No angry mountain lions, no crashing planes, no visible threats of any kind are required: The fear response appears by itself, out of the blue. Those who have survived this kind of event truly understand the full overwhelming impact of fear run amok.

IT WAS JUST ANOTHER busy day for Cindy Jacobs, a thirty-year-old housewife and mother from Tallahassee, Florida. She'd just picked up her two-and-a-half-year-old daughter from day care, and was driving back home along with her one-year-old and one-month-old, the cheerful sounds of Raffi, a children's musician, playing on the tape deck. Jacobs' life was full, but she didn't consider it hectic. If anything, she thrived on the challenge of keeping together a household that also included her husband's two teenagers from a previous marriage.

She was about a mile from home, driving past her favorite fruit stand, thinking about what she was going to make for lunch, when she felt a twinge in her chest. It swelled into a tightness midway between her left breast and her clavicle. Then her left arm got tingly and warm. She gasped and fought for air. *Oh my God*, she thought. *I'm having a heart attack*. She didn't know what to do. Sweat erupted from every pore. She was overcome with the intense fear that she was about to die.

Jacobs slowed the car, flipped her turn indicator, checked the rear-view mirror, and pulled off the road into a parking lot. As the car rolled to a stop she rested her forehead on the steering wheel. The two-year-old was singing along to the Raffi tape, completely oblivious.

I'm about to die with my kids right here in the car.

She was paralyzed. Somehow, she managed to move an arm and reach for her cell phone. She dialed her HMO to ask for clearance to call an ambulance. A nurse answered. "I'm having a heart attack!" Jacobs blurted.

On the other end of the line, the nurse chuckled. "Honey," she said, "if you were having a heart attack, you wouldn't be talking to me on the phone."

That's pretty rude, Jacobs thought. *After I'm dead you're going to wish you hadn't said that!*

All the same, hearing the woman's voice began to soothe her fears. As she talked with the nurse, describing where she was and what she was feeling, she felt her symptoms begin to fade. "You're having a panic attack," the nurse said. "You should go down to the emergency room and get some tests done."

Jacobs didn't want to go to the emergency room. She wanted to go home. She'd just survived some kind of pre-heart attack, that much seemed obvious to her, but with a toddler and two babies to feed, and two teenagers due home from school in a few hours, she didn't have time for fooling around with trips to the hospital. She'd tough it out, just like she did with everything else. If she was lucky her heart wouldn't act up again. She put the car into gear and merged back onto the road.

Over the next few weeks she tried not to think about the attack. Then another one struck, this time while she was sitting in church. The symptoms were the same as before: the squeezing in her upper left chest, the numbness in her arm, the feeling of intense fear. Her doctor told her that they were panic attacks, brought on by stress. "Stress? Me?" Jacobs said. "What do I have to be stressed about?"

She went on with her life. Every week or two, an attack would stop her in her tracks. She made the rounds of the city's medical specialists. Did she have a tumor? Diabetes? An aneurysm? Everyone told her she was fine. But she knew there was something profoundly wrong. Every week or two, she faced anew the prospect of dying. When she would feel an attack coming on she would sit, frozen in fear, her muscles tense—all the effects, although she didn't recognize it, of a full sympathetic freeze response.

Jacobs just couldn't believe her doctor's theory that stress was the cause of her problems. "I don't think I was resisting the idea," she says today. "It's just that it wasn't computing. I couldn't associate the stress that he was talking about with what I was feeling."

At last Jacobs let her doctor prescribe her an anti-anxiety drug, Xanax. He told her that when she felt an attack coming on, she should take one, lie down, and see if the medication helped. To her surprise, it did. "It was amazing," Jacobs says. "It helped a lot. It's not like it blocked my thoughts, but it made it so they didn't bother me."

Finally, she was able to understand what her doctor had been telling her all along. She thought about the fact that her husband had survived a struggle with cancer that had only gone into remission six months before her first attack. She thought about how, in the span of two-and-a-half years, she'd gone from zero to five children. And she thought about how her upbringing in a chaotic household with four other kids had taught her to never complain and never acknowledge weakness.

Oh my God, she thought. *I* am *stressed out.*

BIZARRE THOUGH THEY seemed, Cindy Jacobs' attacks were in fact quite typical of panic disorder, an affliction that will affect one in five Americans in the course of their lives. The first attack often strikes out of the blue, with such intensity that it feels like a heart attack. Often times, the person has been under a lot of stress, but for whatever reason hasn't noticed the symptoms. In Jacobs' case, she was very demanding of herself. She considered herself hyper-competent, and wanted others to see her that way, too. She was not a whiner or a complainer. Even as she found herself overwhelmed with having to take care of a large family, she felt that she should be able to do more. She had little inclination to focus on the twinges of discomfort and fatigue that accompany stress.

And then, one day—*POW!* Her body grabbed her by the lapels and gave her a shake. Specifically, it hit her with a full-blown, damn-the-torpedoes-full-speed-ahead activation of the sympathetic nervous system. The seed of the attack began somewhere in the subconscious regions of her brain, and by the time it entered her consciousness it was already a raging typhoon.

"The first time people have a panic attack, they often think that there's something physically wrong with them," says Randi McCabe, an anxiety researcher and clinician at St. Joseph's Healthcare in Hamilton, Ontario. "They might feel heart palpitations, or sense that they're choking, and think they'd better get to a hospital. They may feel that they're going to lose control over some bodily function, like all of a sudden they may feel like they could throw up, or they could pass out, or they could lose control of their bowel or bladder. Sometimes, if the predominant symptom is depersonalization or feelings of unreality, they may feel like they're going crazy. Then they might not want to go tell anybody."

What triggers a panic attack? A leading theory holds that an imbalance of dissolved gases in the blood is the culprit behind many, if not all, first panic attacks. To produce energy, our bodies take oxygen from the atmosphere and use it to oxidize its internal stores of fuel.

The resulting reaction produces excess energy as well as carbon dioxide and water as waste products. If we don't keep breathing in fresh air, the level of oxygen in our blood goes down, and the level of dissolved carbon dioxide goes up. These changes are sensed by receptors in the brain stem, the heart, and the carotid artery, which together help trigger the urge to breathe. Although we have no awareness of it, this monitoring is constantly going on at a subconscious level.

When we're stressed out or anxious—even if we're not aware of it—the sympathetic nervous system induces rapid, shallow breathing. This can have the effect of over-oxygenating the blood, creating excessively low carbon dioxide levels. This imbalance leads to constriction of blood vessels and, ironically, a shortage of oxygen in the brain. Breathing gets faster and shallower; the imbalance of the dissolved gases gets worse. The problem spirals into hyperventilation. The sufferer feels lightheaded, with pain in the chest and a rapid heartbeat, trembling, and numbness or tingling in the mouth or extremities—some of the same symptoms that are generated by a full sympathetic nervous system response.

As Cindy Jacobs was driving along with her children in her car, she was completely taken by surprise by these symptoms. She didn't know what was going on, except that it felt life-threatening. She became terrified. Instead of dampening down her fear, her conscious mind decided that there was indeed something terribly wrong. Her prefrontal cortex sent a signal to the amygdala: DANGER! The amygdala ramped up the sympathetic nervous response. The locus ceruleus dosed the brain with more norepinephrine, which in turn sensitized the amygdala further.

Jacobs' heart raced faster; her pulse pounded harder; the sweat poured out of her body. She could hardly breathe. No wonder she felt like she was going to die. Instead of forming a negative feedback loop with her amygdala, her prefrontal cortex was forming a positive feedback loop. The whole system was spiraling out of control. Jacobs was helpless to stop it. Her body, which she had spent her whole life in control of, was now in control of her.

It's hard for someone who has never experienced a panic attack to understand how awful they are. The fear is very real. I've experienced them several times in my life and can attest that they are profoundly alarming. For me, the worst part was feeling my mind spiraling out of control and having no idea how much worse it would get. I felt like I was just along for the ride.

It's perfectly natural, under the circumstances, to wonder if your brain is going to crack, if you're going to be driven insane by the terror, or if you're going to die. The first time a panic attack strikes, you simply have no context, no basis for predicting what might happen next. The natural order of the universe up until that point—specifically, the sense that we inhabit a mind that we control, and that it will not suddenly and unexpectedly rear up against us and attack us—has been thrown out the window. Anything can happen, and you sense that it will probably be terrible.

Eventually the panic subsides. But the possibility remains that another horrible episode could strike out of the blue at any moment. Having no idea where the attack came from, and desperate to avoid another, sufferers form theories, consciously or subconsciously, about what was to blame. They might start to avoid public places, or tall buildings, or bridges.

"Whatever people associate with panic attacks, they will start to avoid, as they develop anticipatory anxiety about future attacks," says McCabe.

Anywhere that's hard to get away from quickly becomes particularly suspect. In my case, my anxiety was triggered by a fainting spell that happened in the middle of a crowded party. For months after that, any time I found myself in a large gathering of people I became acutely aware that something similar might happen again, and if it did, I might not be able to extricate myself quickly. This awareness set my heart racing, which felt like the beginning of an attack, which made me more anxious. If I didn't move to the fringes, I would feel myself slipping out of control.

I was on the verge of developing agoraphobia, as a large percentage of panic sufferers do. The dictionary defines the condition as "an abnormal fear of crowds and public places," but that makes it sound irrational. I would describe agoraphobia as "the knowledge that venturing out into social situations has the potential to trigger an incapacitating sympathetic nervous response." It's not irrational at all. Avoiding a situation that could trigger an awful attack is a perfectly logical response to a real threat—one that's no less real for the fact that it happens to exist within a person's own nervous system.

Which is not to say there's nothing to be done. In my case, I found that I was able to gain a modicum of control over my fear simply by moving close to an exit, or sitting on the fringe of a gathering where the crowd wasn't so dense. Just knowing that I had some power over my state of mind helped diffuse the anxiety. Each time I socialized, the level of anxiety became less. Eventually I became confident that I could handle the situation. The feedback loop was broken; my cortex had learned to soothe my amygdala. And with that, the panic faded away.

My case was mild. More severe panic attacks require medical help. Cindy Jacobs' doctor helped her break the feedback loop with Xanax, a member of the class of drugs known as benzodiazepines. These work by enhancing the action of an inhibitory neurotransmitter, gamma-aminobutyric acid (GABA). Among other things, this has the effect of raising the amygdala's activation threshold. Once she took it, Jacobs' terrified thoughts no longer had the same stimulating effects on the amygdala, so the amygdala failed to trigger sympathetic arousal. Jacobs could feel the symptoms of panic subsiding, and that eased her conscious fear. Soon the attack was over.

Benzodiazepines are far from perfect. Xanax is a sedative, and highly habit-forming; it's the most abused prescription drug in the United States. What's more, the body quickly adapts to it, so it stops being effective after about four weeks of continuous use. But for Jacobs, just having a bottle of it in her purse gave her confidence that kept anxiety at bay.

There are several other treatments available for panic. The class of antidepressants known as selective serotonin reuptake inhibitors, or SSRIs, are safer than benzodiazepines and are preferred for long-term treatment. Also effective is cognitive behavioral therapy, a talking cure that focuses on modifying how individuals perceive their body's response, so that the feedback loop can be broken at the conscious end of the circuit.

But for many sufferers, like Jacobs, the most important step is simply to recognize what is going on. Only after she was able to accept her panic for what it was did she begin the long battle to remove it from her life. In time, she came to realize that the cycle of fear had served a purpose. Subconsciously, she'd been running up a debt of stress that was becoming unbearable. Fear was her body's way of communicating this to her. It took her a long time to acknowledge the message, but ultimately she did, and was able to make the changes to her life that she needed.

"Once I started to understand what anxiety and stress was, I went on quite a journey," she says. "I needed to learn that the body has an intelligence that is communicated through a variety of ways. If I pay attention, then my body will let me know, 'Hey, it's time to slow down.'"

Cindy Jacobs has long since fought and won her battle with panic. Still, in a part of her mind, she still has a crystalline, flashbulb memory of the spot on the road where she first thought she was having a heart attack. The memory is as clear as another one she has of another spot on that same road. Years before, as a teenager, she'd been driving along there when her car had been struck out of the blue by a drunk driver. The wreck was terrible, and she was lucky to survive. For Jacobs, both spots along the road are as vivid in her memory—and the fear that she felt in the two spots equally real.

ALTHOUGH ITS MECHANISM may seem baffling to the uninitiated, panic disorder is not an odd, isolated psychological phenomenon. It is part of a much larger family of anxiety disorders, one that also includes social phobia, specific phobia, general anxiety disorder, and post-traumatic stress disorder, which together currently affect 40 million U.S. adults, or nearly 15 percent of the population. All are dysfunctions in which the feedback loop between the amygdala and the cortex spirals out of control, wildly exceeding the demands of the threat at hand. Treatment involves breaking the loop with either cognitive behavior therapy or SSRI medications.

To be sure, it isn't only diagnosed anxiety sufferers who let fear grow out of proportion to the circumstances. We all do it. We've all felt a surge of adrenaline when our jet hits a patch of turbulence, even though we rationally know our risk of crashing is less than one in a million. Or lain awake at night, mulling things that we have no control over. Niggling doubts turn into worry, and worry into corrosive stress. Often, the anxiety we feel is worse in itself than the outcome we fear. Sarah Burgard, a researcher at the University of Michigan, examined medical statistics on more than three thousand workers and found that those who worry about losing their jobs suffer worse health than those who've already been fired.

The antidote to excessive fear is to get a grasp on the mechanisms behind our minds' deceptive tendencies. The more accurately we understand about how our fear circuitry works, the better we can predict how we're going to react and take steps to stay in control.

Granted, parsing the hidden syntax of our brains is anything but a simple task. As neuroimaging and other modern research tools lead us to an ever-more detailed understanding of the brain, the picture that's emerging grows ever more complicated and astonishing. In the next chapter we'll dig down one layer deeper into the neurobiology of the fear response and see that, although all fear might feel alike, the body actually has not one, but two separate fear-response channels, one for each of the two major classes of threat. It's possible for a person to be exceptionally brave in the face of one, and anxious or phobic in response to the other. To begin, we'll meet a man who became a national hero for his bravery in the face of mortal danger—but was reduced to a nervous mess by something utterly harmless.

CHAPTER SIX

IN LOVE AND WAR

WHICH IS SCARIER, a Nazi tank or a pretty girl?

The question wasn't an easy one for Audie Murphy. Growing up in Texas during the Depression, Murphy was an extremely shy and withdrawn young man. The sixth of a sharecropper's twelve children, he dropped out of school after the eighth grade to work as a field hand. He could barely read or write. When he was ten years old, his father abandoned the family. To stave off hunger, Murphy hunted for squirrels, rabbits, and other small animals. In reply to a friend who commented on what a good shot he was, he said, "If I don't hit what I shoot at, my family won't eat today." His hunting skills weren't enough. After his mother died, he was forced to put his three youngest siblings in an orphanage.

Given his limited prospects, it's not surprising that Murphy lacked social skills. Indeed, he was almost pathologically shy. Scrawny and just five feet five, he hardly cut a commanding figure. After the war broke out in 1941, he tried to join the Army, but the baby-faced fifteen-year-old was rejected as too young. The following year, armed with a doctored birth certificate, he tried again. The Marines rejected him for his puny stature, but the Army took him. During basic training, Murphy passed out during combat drills, and his officers, figuring his physique was inadequate for the stresses of combat, wanted to assign him to cooking duty. Murphy pressed to serve in a combat unit, though, and after basic training he was assigned to the 3rd Infantry Division and shipped out to North Africa.

It was during the invasion of Sicily that Murphy first saw action. He subsequently took part in the invasion of the Italian mainland, and then of France. Along the way he earned numerous citations and won promotions to sergeant and then lieutenant.

By the beginning of 1945, the war in Europe was in its final stage. Allied forces had swept across France and were on the threshold of the German heartland. But the Nazi regime was prepared to fight on with suicidal ferocity. On the bitterly cold afternoon of January 26, Murphy was in command of a badly depleted company that was hunkered down in a forest near the village of Holtzwihr, France, five miles from the German border. An intense battle the day before had left 102 of the company's 120 men dead or wounded. All the officers except Murphy had been killed.

And the fighting was far from over. As Murphy watched through binoculars, 250 German soldiers, dressed in white camouflage and accompanied by six Tiger tanks, poured out of Holtzwihr and stormed toward the American lines. Ordering his men to fall back, Murphy stayed in a forward position and used a field telephone to call in artillery fire on the attacking Germans. Shells shrieked in from distant batteries, exploding in dense starbursts among the enemy troops, but still the assault pressed on. To Murphy's right, an American tank was hit by a German shell and exploded, leaving a burning hulk. Two of its crew were killed instantly. The rest ran for safety. Nearly surrounded by enemy, Murphy climbed onto the remains of the tank and began firing its heavy .50-caliber machine gun. Two more shells slammed into it. The vehicle was loaded with ammunition and fuel and could have exploded at any moment, but Murphy fought on, enveloped in thick black smoke. German tanks, unable to see him amid the swirling black cloud, rolled past him on either side. For a moment the icy wind blew the smoke clear of Murphy's position, revealing a squad of twelve German soldiers taking cover in a ditch less than a hundred feet away. Murphy spotted them before they saw him. Swiveling his machine gun, he unleashed a burst of fire and killed them all.

Finding its advance thwarted, the enemy infantry wavered and fell back. The Tiger tanks, vulnerable without infantry protection, retreated as well. Out of ammunition, Murphy jumped down from the burning tank and ran back to the forest to rally his men. The tank exploded. Although he was severely wounded in one leg, Murphy led his men on a counterattack that rolled back the German assault.

For his action that day, the Army awarded Murphy the Medal of Honor, adding to the 31 medals he'd already won in 27 months of fighting. Sent back home, he wasn't just a war hero, he was *the* war hero: the most decorated combat veteran of World War II. He was feted in numerous parades; his picture was on the cover of *LIFE* magazine. The actor Jimmy Cagney saw it and invited Murphy out to Hollywood, where he became an actor too, eventually appearing in 44 movies. His autobiography *To Hell and Back* was a bestseller, and he played himself in the film version.

Yet for all his fame, for all the adulation heaped on him, Murphy remained the same intensely bashful young man. Asked to appear on a radio show, he was so overcome with nerves that actors had to read his lines for him. Even as a movie star he remained shy, avoiding opportunities to socialize with his fellow actors and turning bashful around the pretty young women who flocked to him.

It's hard to reconcile the image of a battle-scarred veteran with that of a timid wallflower. Yet when Murphy was called upon to say a few words at his home town's celebration in his honor, he said that he'd rather face a machine-gun nest than have to give a speech. This is the paradox of Audie Murphy: He was both superhumanly brave and incredibly timid. He was both lion and mouse.

THE PARADOX OF Audie Murphy, of course, like all good paradoxes, points to an underlying flaw in our understanding. In folk psychology, we imagine bravery as a kind of mental armor equally impervious to anything that might make us afraid. It's hard to conceive of a kind of courage that can hold up in the face of a deadly Nazi attack yet crumble when asked to speak to an adoring hometown crowd. Millions of years of evolution, however, have provided us with a fear-response system that recognizes two separate categories of threat. It's possible to be, as Murphy was, highly resilient to one type and yet extremely vulnerable to the other.

So far, as we've discussed the kinds of threats that can trigger our fear-response system, we've mainly talked about acute physical danger, the kind of threat that all animals are programmed to respond to. An attacking predator, an advancing landslide, or a raging herd of buffalo is a hazard that must be avoided immediately. Even amoebas have simple mechanisms for reacting to imminent danger. When Audie Murphy held off a platoon of German tanks, he was suppressing this kind of fear.

But as humans, we also have an additional kind of fear mechanism that many other animals lack. Long ago, before mammals evolved, our early reptilian ancestors were a fairly self-sufficient lot. Mothers laid their eggs and moved on; the infants hatched and started fending for themselves right away. Interpersonal skills were not required. When the earliest mammals appeared on the scene over 200 million years ago, biology reorganized their familial ties. Babies came into the world helpless. In order to survive, they needed warmth, nourishment, and protection from their parents.

That's as true today for a human being as it is for a mouse or an echidna. For a mammalian infant, the most imminent danger is not being done in by an attacker, but simply being left alone. Notwithstanding the ancient Roman myth of Romulus and Remus, a baby left unattended in the wilderness will die. Even as they grow older, mammals remain relatively dependent on fellow members of their species. Compared to reptiles, they are social creatures. Though some species are loners, the majority live in bands (or packs, or pods, or troops, or prides), and rely on one another for protection, grooming, food, and all the other necessities of life. No man is an island, and neither is a baboon or a groundhog.

In order to live in a group, social animals need a way to smooth over the endless array of potential conflicts and compromises that could arise when multiple individuals have to share limited space and resources. To do this, they instinctively create a hierarchy. Chickens create a very simple one, a "pecking order" in which an alpha chicken dominates all the others, the beta chicken dominates everyone but the alpha chicken, the gamma chicken dominates everyone but the alpha and beta chicken, and so on, all the way down to the lowest-ranking chicken, who is dominated by everyone else (and consequently tends to be stressed-out and miserable).

Great apes, including human beings, have a much more complicated social network, one that includes things like shifting alliances and kinship bonds. But the principle is the same: Any time new individuals come into the group, they have to find their place within it, a procedure that involves challenging others and either submitting to them or dominating them. This process is so deeply ingrained that it goes on subconsciously, through gesture and posture and tone of voice. Rituals like saying hello, shaking hands, asking "How are you?," and hugging good-bye are all actions that, though we carry them out consciously, are reflections of an ancient subconscious system for maintaining our social position.

One main brain chemical organizes our social bonding process. Oxytocin is a hormone and neurotransmitter released during sexual climax, childbirth, breast-feeding, and even, to a lesser extent, by the simple act of touching. Oxytocin helps forge the bond between friends, between lovers, and between parents and children. Studies have found that if rats and mice are given drugs that block the effects of oxytocin, they stop taking care of their offspring and are unable to recognize familiar individuals of their own species. Conversely, when members of a loving couple hug each other, their level of oxytocin goes up and their level of cortisol goes down.

Oxytocin plays a key role in the maintenance of the social structure. C. Sue Carter, a researcher at the University of Maryland, conducted a famous study of two closely related species of vole, a small burrowing mammal. Montane voles are closely related to prairie voles, differing mainly in their parenting and attachment styles. Prairie voles have abundant receptors for oxytocin. Males and females form long-lasting bonds, and both take part in child-rearing. Montane voles have few receptors for oxytocin, and the males are promiscuous and do not contribute to child-rearing. What's more, if male prairie voles are given drugs that block oxytocin, they, too, adopt a promiscuous lifestyle.

As social animals, our place in the hierarchy is of vital importance. Our survival is at stake. When our status is threatened, our brains are wired to respond much as they do to a physical threat. The system kicks in from the moment we enter the world. A newborn's cries don't signal discomfort or fear per se; they're cues to induce support from their closest allies in their social network, their parents. Oxytocin is crucial to the instinct. Giving oxytocin to newborn laboratory animals that have been separated from their parents sharply reduces the number of distress calls that they make.

When we're excluded from the support of a social network, cut off from loving physical content, we feel lonely and miserable. The emotional pain that we experience feels like physical pain, and for good reason. Areas in the brain that take part in the social circuit evolved from nearby areas that mediate physical pain. We're not getting metaphorical when we talk about the pain of a broken heart—it physically hurts. Naomi Eisenberger, a neuroscientist at University of California, Los Angeles, scanned the brains of subjects who were playing a ball-tossing game on a computer. As the game went on, the subjects were left out of the ball-tossing between other players, which made them feel excluded. Eisenberger found that their feelings of rejection activated the anterior cingulate cortex, a region also responsible for generating sensations of physical pain.

Just as social rejection triggers pain, the prospect of a social threat triggers the fear response. When you see a stranger's face for the first time, your amygdala instantly activates, as I learned when I took part in Lilly Mujica-Parodi's fear study. It's as though when meeting a new person we're wired to expect the worst. From an evolutionary standpoint, this makes sense. Every time we enter a novel social situation we're submitting ourselves for evaluation. Given the potential for social conflict to cause us pain, we're taking a real risk. And so we have every reason to be fearful. The stakes are high.

This is the kind of fear that Audie Murphy had a hard time managing. Whether through genetic predisposition, as a result of his impoverished upbringing, or a combination of both, his social circuitry was much more sensitive to potential threats than was his circuitry for handling physical danger. For him, a pretty girl really was scarier than a Nazi tank.

MOST OF US CAN RELATE TO how Murphy felt, in kind if not in degree. Social fear is a universal part of the human experience. Forty percent of Americans list public speaking as their number-one fear. And who hasn't felt a wave of trepidation at asking someone out on a date, or speaking up in a public forum, or walking through the door into a party? The involuntary signals that we give off in such situations, like blushing and stammering, are part of our innate set of automatic skills that help us find our place in the hierarchy.

Social fear can also manifest itself in odder ways. Hardly a man hasn't experienced the sudden inexplicable inability to urinate when someone's standing at the next urinal in a public restroom. This well-documented psychological phenomenon is a type of mild social phobia that researchers term "shy bladder syndrome."

When the fear of social interaction becomes too intense, it can wreak havoc on a person's life. That's when normal social fear edges over into the pathological condition of social phobia. People with this disorder suffer from a profound dread of meeting other people. They're withdrawn in unfamiliar social settings and go to great lengths to avoid speaking in front of a group. Their fear can manifest as blushing, palpitations, clammy palms, nausea, stammering, and trembling—or worse. Being the focus of group attention is enough to send their dread spiraling out of control until it becomes a full-blown panic attack.

Agoraphobics also dread public spaces, but the root of their fear is quite different. Their underlying dysfunction is panic disorder, the fear of fear itself. Agoraphobics avoid public spaces because they've come to identify them as places where a panic attack might strike. "Social phobics" are not afraid of panic per se. They fear how others will perceive them. The process of evaluation inherent in every new social situation becomes the occasion for unfathomable dread.

Social phobics fear the worst, imagining that others will see them as stupid, disgusting, boring, or loathsome. And this fear results in heightened sensitivity to cues that reinforce the negative expectation. "People with social phobia are more likely to misinterpret ambiguous signals as negative or threatening," says anxiety researcher Randi McCabe. "So if you don't say 'hi' to me, it's because you don't like me, not because maybe you were busy."

This is the primal social-status circuitry lurching into overdrive. Once the social fear circuitry is triggered, the prophecy can become self-fulfilling. Indeed, social phobics are particularly prone to paradoxical failure, or what psychologists refer to as "ironic effects." The harder they try to control their fear, the worse their dysfunction gets. Explains McCabe: "People with social phobia often take an observer perspective in an interaction. Instead of just taking part in a conversation, they're observing it at the same time. And because they're evaluating and monitoring the interaction, they're not participating fully, and so they're not doing as well. They're sending out negative signals like not making eye contact. And so they do put people off."

Worrying about the problem makes them focus on it more, and focusing on the problem only intensifies it. It's like when you're behind the wheel of a car skidding toward a ditch on an icy road: If you concentrate on the thing you're trying to avoid, the chances are that you'll drive right into it.

A particularly torturous variety of ironic self-defeat happens in the bedroom. Worried about whether they'll be able to perform sexually or not, men can get caught up in the observer perspective, distractedly watching themselves anxiously to see whether their worst fears will be realized. Masters and Johnson called this kind of behavior "spectatoring." Instead of focusing on their partner and allowing themselves to get caught up in the passion of the moment, they reserve a part of their attention for assessing how well the occasion is going. "Does she seem to be enjoying what I'm doing?" he asks himself. "Is she noticing that my erection has flagged? If so, what does she think of me?" Under the circumstances, impotence is all but inevitable.

At present, the recommended treatment for social phobia is the same as for the other anxiety disorders: antidepressants or cognitive behavioral therapy. Compared to their great success in treating panic disorder, however, these therapies are not very effective at treating social phobia. Some 40 percent of patients don't respond to either treatment.

A better clinical approach might be to make use of the hormone at the root of social bonding: oxytocin. Peter Kirsch, a researcher at the National Institute of Mental Health, scanned the brains of volunteers looking at frightening images like angry faces and guns. Their amygdalas showed strong activation; but after the subjects took a few whiffs of oxytocin, that activation decreased. The effect was especially strong when the pictures were of faces. Kirsch concluded that oxytocin is particularly effective at dampening social fear.

Other investigators are working on a drug that would mimic oxytocin's effects on the brain. Simply ingesting oxytocin by itself likely wouldn't be very effective, since it wears off quickly. The ideal solution would be a substance that interacts with receptors in the same way as oxytocin, but lingers in the system longer before being broken down.

You don't have to wait for the future to take advantage of oxytocin's benefits, however. You already have access to an endless supply. In general, it's a bad idea to self-medicate for a psychiatric condition, but in the case of social anxiety, you can safely generate as much oxytocin as you want, right inside your own body. As we've already seen, simply getting a hug from your partner can boost oxytocin and lower stress. Even more effective: sex. A 2006 study found that "penetrative intercourse" reduces the amount of stress people feel before public speaking. The study found that the more sex subjects had, the less stress they felt.

WHEN AUDIE MURPHY returned to America, he didn't seem like he was someone who had anything to be anxious about. Across the nation, he was lionized as the embodiment of the courage that had made victory possible.

His personal life, however, was less rosy than the public imagined. More than two years of brutal killing had inflicted profound damage on a psyche already vulnerable to the self-reinforcing spirals of anxiety. "If you have one anxiety disorder, you're going to be more at risk for developing another one," says McCabe.

In Murphy's day, the psychic toll wreaked by combat was known as battle fatigue, and its mechanisms were poorly understood. Today, we know that post-traumatic stress disorder (PTSD) is a form of anxiety that's caused by learned hypersensitivity to threat cues that have been experienced in the course of real-life trauma. As in all anxiety disorders, the brain overreacts to what it perceives subconsciously as threat cues, resulting in an excessive fear response. In Murphy's case, he suffered from intense flashbacks to his experiences in combat, terrifying images that could be triggered by things as innocuous as, for instance, a bowl of black-eyed peas. Such terrors punctuated long bouts of depression and insomnia. He became addicted to sleeping pills and tried to recover by locking himself in a motel room for a week and going cold turkey.

At the time, it was considered unseemly for veterans to talk about mental-health issues, but Murphy spoke out about his problems, and called on the government to do more to help men returning from war. Even Murphy, however, was unable to get all the help he needed. The trauma he'd suffered haunted him throughout the rest of his life. His marriages broke up, he was arrested for violence, and ultimately died young, in a plane crash, at the age of forty-four.

Murphy never recognized that he suffered from social phobia. The condition wasn't formally recognized in the *Diagnostic and Statistical Manual of Mental Disorders* (DSM), the handbook of psychiatry, until 1980, nine years after his death. Even if the diagnosis had existed during Murphy's lifetime, he still might never have received treatment. Although social phobia is the third-most common psychiatric condition in this country, affecting one in eight Americans in the course of their lifetime, many fail to receive help. Part of the problem is that, in order to seek treatment, social phobics must battle against the very dread of strangers that they find so crippling.

Another factor is that the condition is so often misconstrued. Social phobia can manifest itself in many different forms, which often bear little outward resemblance to one another. We've already encountered urinary blockage in the men's room and sexual frustration in the bedroom. As we'll see in the next chapter, there's an equally widespread form that is liable to be witnessed by millions of Americans on their TV screens every week. Few would guess that what they're seeing is a form of social phobia. In fact, few would even suspect that it's a kind of fear.

CHAPTER SEVEN

"ONE OF THE BIGGEST CHOKERS OF ALL TIME"

ALL STRESS ULTIMATELY comes from within. We may face challenges from the outside world, but it's not until our minds assess a situation and identify a threat that our defense mechanisms swing into action. As we've seen time and again, our physiological response is, oftentimes, the worst problem we face. That's particularly true when the challenge is to earn the social approval of others. When all eyes are on us, the need to perform can have the most crushing, cruelly paradoxical effects. When this kind of trouble strikes, it doesn't just trip us up; it crushes us.

No one knows the bitter sting of such self-defeat better than Dan Jansen. One of the strongest speed skaters that the United States has ever produced, Jansen seemed destined for greatness from an early age. The youngest of nine children in a family that was obsessed with the sport, he took his first skating lessons at the age of four and competed in his first national speed skating championship at eleven. At the age of sixteen, he set a junior world record. Tall and muscular, with a chiseled jaw and movie-star good looks, Jansen was perfectly positioned to be America's boy hero on the ice.

Speed skating is a game of power, balance, rhythm, and precision, especially at the shortest distances, the lengths in which Jansen competed. Two racers at a time start side-by-side and skate counterclockwise around a hockey rink-sized oval. The 500 meter was Jansen's specialty. In less than 40 seconds, a skater must deploy explosive force in no more than 70 or 80 strokes of his skates before crossing the finish line. With only a few hundredths of a second separating a medalist from an also-ran, there is no margin for error.

Going into the 1988 Winter Olympics at Calgary, Alberta, Jansen was the front-runner by a wide margin. At 22, he was in peak physical form, carrying 190 pounds of lean muscle on a 6-foot frame. In 1986 he'd won the World Cup in both the 500 and 1,000 meters. Just a week earlier, he'd taken first place in the World Sprint Championships. He was "The Fastest Man on Ice."

But on this day his focus was off. For more than a year his twenty-seven-year-old older sister Jane had been suffering from leukemia, and lately her condition had begun to deteriorate. On the morning of his first race, his mother called from his sister's hospital room at six in the morning: Jane had taken a marked turn for the worse. Hooked up to a respirator, she was unable to talk, but one of Dan's brothers held the phone up to her ear so that he could tell her he loved her. That afternoon, he received the message that Jane had died. Her last message for him: Win the gold medal.

He stepped out onto the ice at 5:20 p.m. He had drawn the favorable position, on the inside of the track, next to Japanese skater Yasushi Kuroiwa. But he was shaken. After the starter called "Ready!" he jumped before the pistol went off. At the restart, he held back a fraction of a second, then launched into his powerful stride. But something didn't feel right. A hundred yards into the race, at the first turn, Jansen's left skate slipped. He reached out an arm to try to save himself, but it was too late. He sailed into Kuroiwa, knocking him over, and flew into the padded sideboards so hard that he bounced off and wound up back on his feet.

The second race, the 1,000 meter, was four days later. Jansen tried to compose himself. He knew that if he could get back in the competitive mindset he had the physical capacity to beat anyone in the world. This time he got off to a strong start, but as he came down a straightaway midway through the race his skates betrayed him again. He hit the ice and slid all the way to the padded wall of the rink. As he slowly pushed himself upright, 90 million viewers around the world saw on his face a look of utter dejection. For all his past achievements, Jansen knew that the world saw him only as the man who fell at the Olympics.

Over the next four years, Jansen applied himself with absolute intensity, training hard, as he told one reporter, to convince himself "that it wasn't going to happen again." He dominated the World Cup in the 500 meters, winning first place three years in a row. For the 1992 Winter Games in Albertville, France, he was in even better shape than he'd been at Calgary. A mere month before the games, he'd broken the 500-meter world record. But something about the Olympics seemed to have gotten to him. In the 500 meter, his strongest event, he lost his rhythm on the final turn and came in fourth. In the 1,000 meter, he placed an embarrassing twenty-sixth.

Once again he returned to practicing with redoubled ferocity. Due to a change in the scheduling of the Winter Olympics, the next games would be held at Lillehammer, Norway in 1994. Continuing his World Cup domination, Jansen broke the world record in March of 1993, then nine months later broke it again, becoming the first person to ever skate the 500 in less than 36 seconds. It was a feat that some compared to breaking the 4-minute mile in track. That December, *Sports Illustrated* wrote: "[W]ith Lillehammer only two months away, Jansen is skating as never before."

Jansen was twenty-eight years old going into his fourth Olympiad. He and everyone else realized that this would be his last shot at an Olympic medal. Despite his past jinxes, he was again favored to win gold in the 500 meter. Before the event, his coach, Peter Mueller, told the press: "He's the greatest sprinter in the history of speed skating, and he's going to prove it."

On February 14, 1994, six years to the day after his first fall in Calgary, Jansen came out strong from the starting line, powering his way to a world-record pace on the first 100 meters. His technique was flawless until he entered the final turn, 24 seconds into the race. The slip was small—an almost imperceptible burble of his left skate. He swiped the ice with his hand and kept himself upright. But the mistake was enough to shave precious milliseconds off his time. As he crossed the finish line, he ripped off his hood and raised his hands in a gesture of exasperation and disbelief. He had finished in eighth place.

The last Olympic race of his career, the 1,000 meter, was scheduled in four days. He would have one last chance to win a medal; one last chance to redeem ten years of letdown, to claim the prize that everyone believed his talent deserved. In four days, he would either win or lose. He would have the rest of his life to think about what he had done.

By now no one expected Jansen to win. The odds-on favorite was Igor Zhelezovsky of Belarus. Jansen simply lacked the mettle to compete in the world's greatest sporting competition. "This may be the greatest speed skater that ever put on skates never to have won an Olympic medal," sports psychologist Jim Loehr told ESPN. And that, said Loehr, meant "that he was one of the biggest chokers of all time."

CHOKING. IT'S THE MOST DREADED word in sports. To try and fail is one thing—it's an inevitable part of athletic competition. But choking—the sudden, dramatic loss of athletic skill under pressure—is something much worse, an implosion as inexplicable as it is humiliating.

Every sport has its variation of choking. Golfers dread "the yips"—the abrupt inability to sink even the easiest putt. In archery, "target panic" causes technique to unravel so that even the most skilled bowman is unable to release the arrow at the right time. Baseball fans talk about "Steve Blass disease," so called after the Pittsburgh Pirates pitcher—a former All Star and World Series champ—who, in the middle of the 1972 season, suddenly couldn't throw the ball over the plate.

"You have no idea how frustrating it is," Blass told reporters before the start of the following season. "You know you're embarrassing yourself, but you can't do anything about it. You're helpless—totally afraid and helpless."

Though the skills in each sport are different, the way that choking affects the player follows some striking patterns. Typically, it affects the highly skilled player, not the beginner. It strikes in the middle of a career. And, unlike a simple case of the nerves, or jitters, it doesn't gradually fade over time. One of the most horrifying aspects of choking is that, once it has struck, it can strike again at any time. It's like a chronic, debilitating disease. As noted British golf writer Henry Longhurst said of the yips, "Once you've had 'em, you've got 'em."

Players' attempts to cure their affliction tend to make the situation worse. Mackey Sasser was a talented catcher with the New York Mets who suddenly, after three years on the team, found that he was unable to throw the ball in his normal way back to the mound after each pitch—hardly the most challenging athletic maneuver in the game. To get the ball back to the pitcher at all, he improvised a style that involved tossing the ball into his glove several times, leaning far back on his heels, and then launching the ball in a high lob, which took so long to reach the pitcher that runners were often able to steal bases in the meantime. Not surprisingly, he was soon cut from the roster.

What makes choking particularly mortifying is that it's perceived as resulting from a lack of character—a failure of self-confidence, a deficit of will, or a lack of fighting spirit. After Dan Jansen choked at the Calgary Olympics, writer Alexander Wolff speculated in *Sports Illustrated* that he was simply too nice of a guy—that his eagerness to please was interfering with his ability to build up the necessary head of anger that would power him through his difficulties. "After each of his previous failures, he seemed pained and philosophical—anything but mad," Wolff clucked. "Jansen has been too diligent, too sympathetic, too deserving of finally putting all this torture to rest, to make a persuasive angry avenger."

But choking has nothing to do with character. Indeed, one of the things that makes choking so pernicious and so paradoxically resistant to cure is that its underlying mechanism is so different from what most people commonly assume. The truth? Though its manifestations are quite different from blushing and stammering, choking is in fact a form of social phobia—in psychological terminology, "nongeneralized social anxiety disorder." A clue to its true nature is that it only happens when other people are watching. In practice, Steve Blass threw pitches with the same bullet-train speed as ever. Mackey Sasser threw normally when the stands were empty. Dan Jansen never slipped when he was warming up, or even when he was skating for the title in front of the relatively sparse crowds at the World Cup. It's not the sport that causes choking; it's the spectators.

As we saw in the last chapter, human beings have elaborate psychological equipment that allows them to assess and adapt to social structure. While oxytocin is a key hormone in forming bonds, when it comes to competition, a more important substance is testosterone, the hormone of dominance and male sexuality. In most mammalian social structures, males compete for access to mates, and testosterone alters their physical development to let them do so. In humans, it makes the voice deeper, the beard shaggier, and the muscles bigger. In elk, it makes the antlers grow larger so that an individual can battle for control of a herd of cows. Biologists have found that after a dominance battle, the male who wins experiences a surge of testosterone, while the loser suffers a decline.

Human males don't lock horns, but we compete in other ways. One of them is on the playing field. Victorious sports teams are rewarded with elevated levels of testosterone in their blood, especially after they defeat a longstanding rival. This phenomenon gives rise to the so-called winner's effect: When a victor's testosterone surge jolts his confidence and risk-taking, it gives him a better chance of winning the next time. Success breeds success. A similar effect has also been seen in the workplace. Cambridge University psychologist John Coates studied derivatives traders in the City of London and found that those with higher testosterone levels made, on average, more profitable trades.

"Winning isn't everything," as the saying goes—but as far as the subconscious in concerned, it's worth an awful lot. Sports might be play, but the stakes are real. Social standing and access to mates are crucial factors in determining whether one's genes will be passed along to the next generation. In evolutionary terms, being perceived by others as a winner is a matter of genetic survival. And, as we've seen, social threat triggers social fear. And social fear can yield cruel paradoxical effects.

Choking manifests itself differently from generalized social phobia. It doesn't feel like fear, for one thing. But like other kinds of social fear, the underlying cause is the sufferer's own self-awareness. Instead of being directed outward, the attention is directed inward. Rather than focusing on the task at hand, the question becomes, in the words of former New York City mayor Ed Koch: "How'm I doing?"

The more time a player has to contemplate his plight, the worse it gets. In baseball, pitchers and catchers are particularly vulnerable, because they have all the time in the world to make their throws. When there's no time to think about what they're doing, there's no problem. Mackey Sasser was easily able to throw out base runners as a play unfolded spontaneously—the action happened so quickly that he didn't have time to choke.

Some writers have described choking as "thinking too much," but if it were that simple, then battling it would be a straightforward matter. In fact, choking is far subtler and nefarious than just overthinking. Like other manifestations of anxiety, choking is a paradoxical effect that maddeningly resists any efforts to fight it—generally, it only gets worse if you try to tackle it directly. But by understanding how it works, we can figure out how to outwit it.

AS WE'VE SEEN, WELL-LEARNED motor patterns are nearly as fast and reliable as inborn reflexes. And they're very resilient to stress. If you've practiced drawing and shooting your pistol ten thousand times on the shooting range, the chances are good that you'll also be able to draw and shoot flawlessly when an attacker is running at you with a raised hatchet.

That rule goes out the window, though, when the stress is social in nature. In an attack of acute self-consciousness, even exceptionally well-learned motor patterns go haywire. Dan Jansen had skated tens of thousands of times around rinks like the ones he stumbled on at the Olympics. Steve Blass and Mackey Sasser could have thrown baseballs between the mound and the plate in their sleep. The absolute flawless-ness of their automatic processes didn't do them a bit of good.

So what happened to those well-honed automatic processes? Turns out they functioned just fine. Unfortunately, something else was hap-pening as well.

An elegant experiment into the mechanism of choking was con-ducted by psychologist Rob Gray at Arizona State University. He set up a virtual-reality batting cage for two groups of subjects, one con-sisting of experts, the other of beginners. As they tried to hit pitches in the simulator, a high- or low-frequency tone would sound at a random point during their swing. At first, Gray asked the batters just to ignore the tone—this was the control condition. Later in the experiment, he asked the subjects to listen to whether the frequency of the tone was high or low. Finally, he asked the subjects to determine whether, at the moment they hit the ball, their bat was moving upward or downward.

Gray found that in the third part of the test, when the hitters were paying attention to their swing, novice batters did better than when they were ignoring the tone and simply swinging at the ball. That's not surprising: Beginners carry out tasks by paying attention to what they're doing, so they benefit from the extra focus. Conversely, they did the worst when they were trying to identify the pitch of the tone, as that task distracted them from concentrating on their swing.

For the expert batters, however, the story was almost exactly reversed. Listening to the tone didn't hurt the expert batters—they did just as well as when they were ignoring it, since their swings were car-ried out by automatic processes. However, when they had to report on whether their bat was going up or down, their performance suffered. Numerous other studies have found a similar effect: When expert per-formers start paying attention to what they're doing, they undermine the automaticity that underlies their expertise.

The stress of competition only makes matters worse. Daniel Gucciardi and James Dimmock, researchers at the University of Western Australia, performed a similar experiment with expert golf-ers, and this time added an element of pressure. Their subjects were asked to putt while focusing either on three words that related to their technique—such as "arm," "head," and "weight"—or on three words that had nothing to do with putting—for instance, the names of colors. Then they performed the same routine in a high-anxiety situation, in which whoever putted best would receive a cash reward.

Gucciardi and Dimmock found that when the golfers were thinking about irrelevant words in the high-anxiety setting, their performance didn't suffer. Neither the verbal distraction nor the pressure hurt their putting. But when they were cued to pay attention to their swing, the quality of their putting declined. Bottom line: Self-consciousness plus pressure equals bad news. Substitute social phobia for self-consciousness, and Olympic finals for pressure, and you've got the recipe for a Dan Jansen-scale disaster.

IF CHOKING MEANS thinking too much, then the answer is to think less. Unfortunately, that's just about impossible. Such is the paradox of anxiety. Once you know the potential is there—once you've had your first panic attack or your first choke—you can't un-know it. Every time you go back to the situation where you came to grief before, it's all you can think about. Dan Jansen on the 500-meter starting line: *choke, choke, choke, choke.* Mackey Sasser squatting behind home plate: *choke, choke, choke, choke.* Steve Blass on the pitcher's mound: *choke, choke, choke, choke.*

Trying not to think about it will likely only make the problem worse. Harvard psychologist Daniel Wegman has written extensively about paradoxical mental effects—in particular, the difficulty of trying *not* to think of something. If someone tells you not to think of a white bear, for instance, you can try for a while, but it's bound to pop up into your consciousness sooner or later—and probably sooner. Wegman postulates that when we try to perform a mental task, another part of our brain, which he describes as "the automatic process whereby we monitor control failures," is meanwhile checking in from time to time to see how we're doing. As we're trying not to think of the white bear, then, the monitoring process pops its head through the door to ask: "How's the not-thinking-of-a-white-bear coming along, then?" Oops: white bear.

Of course, it's easy to avoid thinking about a white bear until some-one suggests the idea in the first place. In the same way, you'll never worry about choking until the day you first experience a choke—then, as Henry Longhurst said, "you've got it." In many cases, you'll never know what set off the affliction in the first place. For Sasser, the prob-lem appeared out of the blue after he was hit on the shoulder during a game. It wasn't a serious injury, but from then on he was dogged by his loopy throw back to the mound.

Once a player has the terrible awareness of choking in his head, there's no easy way to get rid of it. The more we try to ignore what we're doing with our body, the more our attention lingers over it. Just as a person suffering from a panic attack feels his heart racing and won-ders if he's about to have another attack, the choker notices his body doing the wrong thing and thereby sets up an unwanted feedback loop, amplifying the symptoms until his expertise is tied up in knots.

Still, the phenomenon seems baffling. How can thinking disrupt hitting and throwing? The answer has to do with the way that the brain carries out movement. Recall that when we first learn a motor skill, like hitting a forehand in tennis, we have to start by thinking it through consciously, but that with practice the process gets encoded into the ventral striatum, so that we no longer have to consciously think about what we're doing. The crucial thing to realize is that there isn't just one pattern stored in the ventral striatum for each kind of action. Rather, for any given action, there are many different ways to carry it out. When I hit the ball, I could turn my wrist this way, or I could angle it that way. I could swing the racket quickly, or slowly. All the possible options are stored together in the striatum.

When it comes time for the action to be carried out, these differ-ent possibilities compete against one another. The strongest one wins, and causes its encoded action to play out. A signal is sent to a region called the globus pallidus, which in turn is connected to motor pattern generators in the cortex. These then cause the muscles to move.

The connection between the globus pallidus and the motor pat-tern generators are not straightforward, however. Rather than simply telling them "Go!," the globus pallidus is constantly sending inhibitory messages to the cortex, telling the motor pattern generators *not* to fire. Essentially, it's functioning like a series of thousands of tiny brakes, each preventing a different motor pattern from being acted out. In order to get anything done, then, neurons in the ventral striatum have to block the neurons in the globus pallidus that are blocking the action. It's action by *inhibition* of *inhibition*.

This rather complicated system can come unglued when the wrong area in the globus pallidus is inhibited. The result is that the wrong motor pattern generator is disinhibited, and some unwanted action takes place. This is the mechanism behind Tourette's syndrome, the condition whose sufferers are plagued by undesired tics and verbal ejaculations. It isn't that some part of their subconscious wants to carry out these actions, but rather that the mechanism to inhibit them is broken.

Something similar is happening when we choke. In this case, what's causing the inappropriate disinhibition is simply the act of thinking about what we're trying *not* to do. As counterintuitive as it may seem, merely thinking about an action can disinhibit the corresponding motor pattern generator and cause the action to take place. This phenomenon, called the "ideomotor effect," was first investigated back in the late nineteenth century. Subjects asked to hold a small pendulum still and to think about movement in a certain direction often found the pendulum moving in that direction, seemingly of its own accord. As pioneering psychologist William James put it, "every representation of a movement awakens in some degree the actual movement which is its object."

When we choke, intense self-consciousness causes us to focus awareness on the act that we're about to perform. The ideomotor effect causes an unwanted inhibition of the globus pallidus, so unleashing undesired motor patterns. We wind up doing a combination of our well-learned motor skill and another, less-desired action—what sports scientists call "double pull"—with the sum being even worse than either one alone.

Unlike the normal degradation of performance that the Yerkes-Dodson curve predicts, the deterioration happens very quickly. A choker doesn't slowly lose his skills as the pressure builds. Rather, he reaches a threshold of anxiety and suddenly collapses. It's black and white, night and day.

Tim Woodman is a British psychologist who has worked with his colleague, Lew Hardy, in developing what they call the "catastrophe model" of performance decline. "As physiological arousal increases, so does performance, up to a certain point," Woodman says. "Beyond that, further increases in arousal have a serious dramatic effect on performance, and that's what we call 'catastrophe.' You don't have a smooth, slight downfall in performance. You have a big drop."

Central to their theory is the idea of hysteresis, the idea that an athlete's trip up the Yerkes-Dodson curve is different on the way up than it was on the way down. "Here's an example I like to give my students," Woodman says. "Imagine you're trying to study and your next door neighbor is playing his stereo at a fairly low level. Then he starts to crank the volume up very gradually. You can tolerate that music for quite some time, up to a certain point, and then you suddenly just lose it, and you can't concentrate any more and you have to go and tell him to turn it down. And he can't just turn it down 10 percent; he's got to turn it down a long way before you can start studying again."

Likewise, once an athlete starts to choke, there's no way he can just stop and take a breather, settle down a bit, and hope to get his mojo back. "To recover good performance, you can't just take it back a notch," says Woodman. "You've got to take it back ten notches before you can start to re-compose yourself."

Archery coaches make use of this fact when they help their students overcome target panic. Typically, an afflicted archer either releases the minute she sees the bull's-eye, or freezes and can't release the arrow at all. One of the most oft-used remedies is to remove the target altogether and let the archer shoot at a bale of hay. It's an effective way of taking the pressure off, thereby resetting the hysteresis curve back to the beginning. After the archer becomes comfortable again in a stress-free setting, she can then begin a positive march back up the Yerkes-Dodson curve.

Another useful approach for helping choke-prone athletes avoid turning their focus inward is to instead consciously fill their minds with productive thoughts. Gucciardi and Dimmock's study found that the golfers who performed best were experts who repeated to themselves a single word that represented a successful putting action, such as "smooth."

"What you really want is what we a call 'holistic process goal,'" says Woodman. "You want to focus on the whole feeling of the performance, rather than on a specific element of it. If I'm golfing, I don't want to focus on the angle of my wrist or how square my shoulders are. I want to focus on whatever feeling it is that you have when you hit a sweet ball."

TO GET HIMSELF READY for the 1994 Olympics, Dan Jansen worked on such a holistic strategy with sports psychologist Jim Loehr. Together they worked out a series of visualization techniques that he would use to steel himself before each competition. But the night before his final race at Lillehammer, Jansen didn't bother to use them.

He was done trying. He'd beaten his head against the wall too many times. Time and time again he'd disappointed himself, disappointed his family, and most of all, disappointed the millions of fans who'd cheered him on and offered encouragement after each of his past failures. He knew that he had the physical strength to deliver, but something in his character seemed determined to undermine all his best efforts. *I'm supposed to win*, he told himself, *but something always goes wrong and they can't celebrate*. He called a reporter at his local newspaper and asked him to pass along a message to his fans: "Sorry, Milwaukee."

He was too tired and fed up to go through the public dissection again. Wearily presenting himself before a press conference, he told reporters, "Life goes on. I have a wife and a child to go home to. I'll just have to live my life without an Olympic 500-meter gold medal." As for the 1,000 meter, he was just going to skate and get it over with. "If it happens, it does," he told the reporters. "If not, I'll go on—same old thing."

Still the letters, phone calls, and faxes poured in. Jimmy Buffet sent a single-page fax on which he had scrawled, simply: "Dan, Blow the Volcano!" But Jansen had no intention of doing any such thing. He didn't feel well, and, at any rate, he'd seen time and again how that kind of desperate intensity had backfired. He was just going to skate.

The Olympic finals in the 1,000 meter event took place on Friday, February 18. Jansen came strong out of the gate, covering the first 200 meters in a mere 16.71 seconds. At the 400-meter mark, where he had to swap lanes with racer Junichi Inoue of Japan, he was able to catch a draft off his competitor, gaining a crucial few hundredths of a second. By the 600-meter mark he was on a world-record pace.

But the legacy of his Olympic demons was still hanging over him. Perhaps by now dark thoughts had started to intrude into the total focus that had carried him through from the start. Sure enough, on the second-to-last turn Jansen lost his rhythm. His foot slipped, his arm swung down in an attempt at a save. His fingers barely grazed the ice as he regained his balance. It had cost him—a hundredth of a second or two perhaps—but he was still in the race. He had survived.

Jansen poured it on. His thighs screaming from fatigue, he swung his skates hard through the final 200 meters. As he crossed the line, his time flashed on the big electronic screen: 1:12.43. This time, his hands went to his head not in despair, but elation. He had bested the world record by 11 one-hundredths of a second—and earned himself a gold medal.

"It's all over," his tearful wife told television cameras. "The past is all behind us now. Oh God!"

After his victory, Jansen became a hero on a scale that he never would have achieved had he simply won his gold medals as expected. Ronald Reagan, George H. W. Bush, and Bill Clinton sent personal congratulations. He was inducted into the United States Olympic Hall of Fame, given the Sullivan Award by the Amateur Athletic Union, and appeared on the cover of *Sports Illustrated*. By turning in a world-beating performance in the 1,000 meters, Jansen had done something much more difficult than besting his rivals: He'd triumphed over his inner demon.

Everyone understands how to skate fast, even if they can't do it themselves. But to overcome a six-year choking spell is a rarer and more mystifying feat. Few could have guessed the secret of his success: In his darkest hour, Jansen gave up hope, and in so doing managed to reset his hysteresis curve back to the beginning.

Although Jansen retired from competitive skating after Lillehammer, he remains in the public eye, working as a commentator on NBC Sports and as a skating coach for the Chicago Blackhawks hockey team.

And although he no longer has to worry about choking, his new life in the front of the camera could set him up for yet another variety of social fear—one that torpedoes not physical, but mental performance. Given the protean nature of social phobia, it should come as no surprise by now that this other form of fear manifests in a strikingly different way from choking. In fact, few suspect that they are two forms of the same condition. Beneath the outward signs, their mechanisms are identical. Instead of leaving us physically helpless, however, this new fear strikes us mute.

CHAPTER EIGHT

THE EYES OF OTHERS

BEING WATCHED TRIGGERS a powerful, primal emotional response, one that's threaded deeply into our evolutionary and cultural history. As primates, we're particularly dependent on vision for learning about the world and for inferring the emotional state of our social group. Like all primates we have, as a corollary, a deep-seated discomfort at being looked at. Brain-scan studies have shown that a direct gaze stimulates the amygdala much more strongly than an indirect gaze. The first features of a face that babies learn to recognize are the eyes, and we are particularly attuned throughout our lives to the odd power possessed by two circles placed side by side: Witness everything from the arc-and-two-dots iconography of the famous yellow smiley face to the eyes painted on the bows of Mediterranean fishing boats to ward off evil spirits.

If the direct gaze of one set of eyes is arousing, being watched by many can be horrific. As noted earlier, the dread of public speaking is all but universal, and a plurality of Americans cite it as their number-one fear. Even seasoned actors feel gut-wrenching terror at the prospect of stepping out into the sea of eyes that awaits them onstage. It's not uncommon to hear of singers and actors who vomit before each performance. There is something about standing in the direct gaze of hundreds or thousands of people that jolts the human psyche.

But for some experienced performers, an even darker phantom lurks—a terror more mysterious, unpredictable, and potentially devastating than the usual pre-show jitters. In its mildest form, it's called "going up": the actor finds herself suddenly blocked, dry, her performance at an unexpected halt. She doesn't know the line, and she doesn't know what to do next. All eyes are on her, and she's lost. If she's lucky, she'll regain her footing and the moment will pass. If she's unlucky, the crisis will be far more severe. Actors can become paralyzed, struck mute on stage as minute after awful minute crawls by, or run panicked from the stage. Careers have been ended with guillotine swiftness by the unexpected pounce of this demon. This is the dreaded specter of stage fright.

No performer is immune, from the first-time novice to the celebrated veteran. If anything, it's the seasoned professional who is more likely to succumb. Sir Laurence Olivier was among the most gifted and protean actors of the twentieth century, a performer whose laurels form a veritable Cobb salad of encomiums, including four Oscars, five Emmys, and a seat in the House of Lords. A titan, in other words. If anyone should have felt safe and self-confident on a stage, it was Olivier.

But no. The blow first fell in 1964, when Olivier was fifty-seven years old and had been treading the footboards for more than four decades. As he described the ordeal in his autobiography, *Confessions of an Actor*, he had had a particularly grueling week. He was not only rehearsing in London for the opening night of Ibsen's *The Master Builder*, in which he played a starring role, but he was also overseeing his company's troubled production in Manchester. In the midst of all this, his wife suffered a miscarriage. Emotionally fraught and severely sleep-deprived, he began to suspect that he did not possess the stamina to pull off his upcoming debut. He had long harbored a fear, he wrote, "that some overblown claim to pride in myself would be bound to find the punishment that it deserved. Such punishment was now served upon me in the form of a much-dreaded terror which was, in fact, nothing other than a merciless attack of stage fright with all its usual shattering symptoms."

Olivier made it through the dress rehearsal, but as the opening night approached he became increasingly anxious that he would not be able to remember his lines. With the opening curtain now mere hours away, his sense of dread continued to mount. "My courage sank, and with each succeeding minute it became less possible to resist this horror," he wrote. But the show must go on. He dressed, applied his makeup, and stood ready in the wings. The curtain went up. At last came the cue for his entrance. Summoning his last shreds of professionalism, Olivier walked out onto the stage, fully convinced that he was about to make a complete fool of himself. The leading lights of British theater were in the crowd, as well as the press. Olivier knew the first few lines, but beyond that he was sure he would be stuck.

He said his lines, took a step forward, and came to a stop. His voice faded, his throat tightened up. It seemed like the room was spinning. This was it: the moment that actors dread. To his dismay, Olivier saw that Noel Coward, one of the most talented showmen in the world and a man of merciless wit, was sitting in the front row. Olivier knew that Coward would perceive his failure right away, and would be scathing in his condemnation. Olivier's humiliation would be total: "It would mean," he wrote, "a mystifying and scandalously sudden retirement."

FOR THOSE OF US who have not experienced the horrors of stage fright, it's difficult to grasp the enormity of such a moment. But the terror is equivalent to that aroused by actual, mortal danger. The sympathetic nervous system launches into full overdrive, generating a physiological response appropriate for a life-or-death crisis. Actors say that "going up" is a sensation a good deal like plummeting from a great height.

"It feels like somebody pushed you out of a plane," says Sidney Fortner, a New York City stage actress. "Just total freefall. And you don't know what's up, nothing's in your mind. It's like waking up in a strange bed and not knowing where you are. You're just totally lost."

Like Olivier, Fortner had been acting for decades before the demon struck. So had Chris Wells, a Los Angeles stage actor who first was afflicted in his early thirties. "It was hideous," he says. "An immediate, instant loss of all awareness and control." The moment occurred while he was playing Lady Bracknell in a production of *The Importance of Being Earnest* at a small theater in Culver City, California. He was midway through a scene when, he says, "I went up in a way I never had before. I was so blanked that I said nothing. And part of what made it so hideous was that it was totally unexplainable. It wasn't as if I'd been skating on thin ice for a while and then broke through. It was just— boom!" Like Sue Yellowtail when she was attacked by the mountain lion, he blacked out. "There was a loss of time, like a moment that was blank," Well says. "I don't even know how long it lasted." The next thing he remembers was being engulfed by a cavernous silence. Finally, his partner in the scene fed him a line, and he was able to continue, but for the rest of the performance he was profoundly shaken.

Like an athlete's first choke, an actor's first attack of stage fright tends to occur in mid-career. Like a panic attack, stage fright often occurs in the wake of other stress in a person's life. And like both, once unleashed it's a demon that from then on lurks in the margins of awareness, always threatening to reappear. One study of symphony and opera musicians found that 24 percent listed stage fright as a primary health concern.

The similarities between choking and stage fright are not coincidental. From a psychological perspective, they are both manifestations of what clinicians call "nongeneralized social anxiety disorder"—also known as performance anxiety. Like a catcher who suddenly can't figure out how to toss a ball back to the pitcher's mound, the actor gripped by stage fright finds himself gripped by a worsening spiral of self-awareness that tramples the automaticity of his expertise.

An experienced actor relies on a suite of techniques learned over the course of many years. Honed to the point of reflex, these can be called upon effortlessly amid the flow of the performance. "You don't have to *know* what's next. It's in your body," says Fortner. To get back on track after going up, she says, she would focus on the props she was using, her location on the stage, and the actions that her character was undertaking: "I would go back to physically what was I doing. Because it was all linked to the physicality, and I knew physically what I had to do next. And when I started doing it, then the line would come."

When Olivier went up during the premiere of *The Master Builder* he used a similar technique to regain his bearings. "I retraced my step," he wrote. "With unusual inaudibility, owing to my tightly clenched teeth, I somehow got on with the play."

The battle against stage fright, then, is the battle to return to automaticity from the appalling condition of self-awareness that lies at the heart of all social anxiety—the feeling of watching oneself, of standing outside the conversation or the performance and assessing oneself as one imagines others are doing. "If you focus on the stage fright, and not on what you're doing," says Fortner, "you can absolutely lose yourself."

Wells once played a role that was so technical that he could perform it without losing himself in the character, as he was accustomed to doing. That left his mind free to wander, and as a result he felt that the risk of stage fright was particularly acute. "It was like my mind split—I was always aware, I was always observing what I was doing," he says, "And that really freaked me out even more, because I had the distinct sensation of: 'Oh my god, now I'm saying this. Am I going to remember the next word?' It was almost like I was leaving the stage and looking down on the proceedings."

Once the panic spiral begins, the arousal of the sympathetic nervous system leads inevitably to the shutdown of higher cognitive functions. Sian Beilock, a psychologist at the University of Chicago, has studied the ways in which high-pressure situations affect the ability to perform cognitive operations. She found that people with more working memory—that is, the ability to manipulate more complex pieces of information in their conscious thoughts—actually did worse when asked to take a math test under a high-pressure situation. She theorizes that people with large working memories become accustomed to relying on complex mental techniques, and that when their working memories shrink under the effects of pressure they're no longer able to use their habitual methods, and thus are left helpless.

A similar effect may explain why experienced performers are more vulnerable to stage fright than novices. Over the years, experienced performers have accumulated a more complex repertoire of techniques. When the anxiety tsunami hits, then, they're left without any tools. That's certainly how it felt to Wells, after his first attack. "I think of myself as a very smart, precise actor, and very technically proficient," he says, "and suddenly I didn't have any weapons. I was defenseless."

THE AUDIENCE MAY NEVER notice that an actor has been seized by an intense attack of stage fright. Hiding his feelings, after all, is an actor's stock in trade. He may collect himself before the silence becomes noticeable and carry on, as Olivier did. But from then on, the question will haunt him like a phantom: Will it happen again?

Some performers conclude that it will, and flee the stage for good. Singer Carly Simon gave up public performance for eight years due to her intense recurring experience of stage fright; pianist Vladimir Horowitz left the stage for fifteen years; singer Barbara Streisand for twenty-seven. Many performers never come back at all. A study of musicians with stage fright found that 30 percent eventually stopped performing or left the profession.

Those who stay often try to deal with the lingering dread via whatever methods they can—a mode of coping that psychologists call "lay interventions." Among performers, good-luck tricks, rituals, and superstitions are rife. Famously, actors say "break a leg" rather than "good luck" before a performance, while opera singers offer "*in bocca al lupo*" ("in the mouth of the wolf"). Ballet dancers consider it good luck to prick a finger and put some blood on a toe shoe before a performance. And it's ancient thespian lore that actors should never refer by name to *Macbeth*, but instead must call it "the Scottish play."

Another type of lay intervention is the improvisational making-do that Mackey Sasser used to get the baseball back to the pitcher's mound. To help him deal with his dread during a run of *The Merchant of Venice,* Olivier asked his fellow actors not to look him in the eye, but instead to deliver their lines to the general vicinity of his face. During a production of *Othello*, Olivier begged one of his fellow actors to remain on stage with him during a soliloquy, so that he wouldn't feel that he was facing the audience alone.

As Mackey Sasser found, such improvisations are only stopgaps at best. After one particularly fraught performance of *Othello*, Olivier felt that he had reached the end of his rope. He sought out some old friends, Lewis Casson and Sybil Thorndike, and asked them what he should do. "Sybil promptly answered, 'Take drugs, darling, we do.'" Olivier didn't bat an eyelid. Soon, he was getting through his nightly performances with the help of Valium, a benzodiazepine.

People have been self-medicating to alleviate fear since the dawn of time. Probably the earliest of the anxiolytics—that is, drugs that reduce fear and anxiety—was alcohol. The phenomenon of "Dutch courage" was well known to the ancient Greeks, who carried wine and drinking cups with them when they marched off to battle. The Celtic tribes of the era went one better, storming into battle not only roaring drunk, but stark naked. The tradition of military tippling was carried on in a more subdued manner by Britain's Royal Navy, which issued its sailors a daily rum ration from the age of sail right until 1970. If any doubt should remain that booze can drown fear as well as sorrow, a 2008 brain-scan experiment found that subjects who were intoxicated up to the legal driving limit showed minimal amygdala activation when shown pictures of frightened faces.

As a bulwark against fear, alcohol has some drawbacks. It impairs judgment, ruins muscular coordination, and tends to make you feel absolutely terrible the next morning. With the advent of the modern chemical industry in the nineteenth century, the path was clear for better substitutes. The first synthetic anxiolytic drug, barbituric acid, was produced in 1864 by the German chemist Adolf von Baeyer. Today the best known of the barbiturates is Seconal, the brand name for secobarbital. While better at calming anxiety than alcohol, barbiturates are also powerfully sedative, highly addictive, and prone to overdose. Seconal turned up in the autopsies of such luminaries as Judy Garland, Jimi Hendrix, and Marilyn Monroe.

As a treatment for anxiety, barbiturates were eventually replaced with benzodiazepines, of which Valium was an early popular variety. While safer than barbiturates, benzodiazepines are also quite addictive, as Olivier discovered to his detriment. By the 1970s, he was taking

Valium every day, along with Mogadon, another benzodiazepine. Among other side effects, the drugs left him in a mental haze that torpedoed his ability to memorize lines.

While the search for the perfect anxiolytic continued, performers stumbled upon a drug that by accident happened to work surprisingly well for stage fright. In the 1950s, Scottish pharmacologist James W. Black synthesized propranolol, a chemical that blocks the action of epinephrine on certain receptors within the sympathetic nervous system. Specifically, it prevents the neurotransmitter from binding to β1- and β2-adrenergic receptors—a fact that earned the drug the designation "beta-blocker." Black was interested in the sympathetic nervous system's effect on the heart, in particular; he wanted to prevent the damaging long-term effects of stress. In that regard, propranolol was a breakthrough success, revolutionizing the treatment of such coronary ailments as abnormal heart rhythms, angina, and high blood pressure.

Along the way, the drug turned out to have an intriguing side effect: It dampens other aspects of the sympathetic nervous response, such as shaking and sweaty hands. This benefit had obvious appeal for classical musicians, whose performance depends on the flawless deployment of fast, intricate motor movements. Although they hadn't been approved for the treatment of performance anxiety, beta-blockers spread like wildfire through orchestra pits and classical music halls. "The little secret in the classical music world—dirty or not—is that the drugs have become nearly ubiquitous," wrote the *New York Times* in 2004.

Because they take effect quickly and their effects last only a short time, beta blockers have gained a following among all sorts of people who are called on to perform, including ordinary people who have to stand up in front of a conference room full of colleagues. The main limitation is that, because the drug prevents blood pressure from increasing, it can't be used in situations that involve significant physical exertion, like dancing or singing. Users don't notice a perceptible effect on their thoughts or emotions; beta-blockers seem to squelch anxiety's feedback loop between the frontal cortex and the amygdala, or even run it in reverse. Seeing that they are showing no outward sign of fear, performers feel more in control, and that notion become self-fulfilling.

As choking is a close kin, psychologically, to stage fright, it's not surprising that beta-blockers have found favor among athletes as well, at least in sports that require precision and steadiness rather than endurance or power. Unlike steroids, which improve an athlete's performance, beta-blockers simply remove the impediment caused by anxiety, allowing a performer to achieve the same level of excellence in competition as in practice. Nonetheless, the international governing bodies of many sports—including archery, riflery, diving, and synchronized swimming—have banned their use in matches. In August 2008, North Korean pistol-shooter Kim Jong-su was stripped of two Olympic medals after a doping test turned up positive for propranolol.

Not everyone is convinced that beta-blockers are the panacea for performance anxiety. One concern is that so many people are taking them without therapeutic supervision. One survey found that 70 percent of classical musicians who were taking beta-blockers obtained them from acquaintances, rather than from doctors. The potential dangers of taking prescription drugs in such an informal way include the possibility of life-threatening interaction; these are powerful cardiac medicines, after all, with potential side effects such as weight gain, nightmares, and depression.

Then there's the issue of how it affects the performance. Critics say that beta-blockers can dull a performance, preventing a musician or actor from reaching the optimal energy at the apex of their Yerkes-Dodson curve. "Anxiety is not all bad," says Linda Hamilton, a former ballet dancer who is now a psychologist specializing in performance issues. "I say to people constantly that there's nothing wrong that you're getting anxious before going onstage, because the fear is gearing you to rise to the occasion. Usually you calm down after the first couple of minutes, and you're just right in the moment, and that's just where you need to be."

Worse, users can come to rely on anxiety medication as a psychological crutch, which, while helping them get by for the time being, may ultimately delay them in seeking effective treatment that could result in a long-term cure. Anxiety researcher Randi McCabe recalls one anxiety patient who for years carried around a bottle containing a single tablet of an anti-anxiety drug. "He'd never taken it, he only had one left, it was almost in crumbs. But he couldn't leave home without it. It was like an American Express card. It made him feel, 'Okay, if I need it, I have it.' But it was like a crutch—a safety behavior." Ultimately, the patient underwent cognitive behavioral therapy and successfully tackled his underlying condition. "He challenged his fears, and built up his confidence in being able to cope with things," says McCabe. "After that, he was on top of the world"—and able to leave his crumbling pill at home.

Cognitive behavioral therapy has been found to be an effective treatment for stage fright, and Hamilton says that most patients see improvement within three months, and sometimes in as little as a few sessions. The underlying principal is that one learns to identify the thought patterns that lead to trouble, and to replace them with more positive cognitions. For instance, Hamilton encourages patients to avoid the kind of nitpicky self-analysis that also plagues chokers and instead to focus on a holistic vision of their performance.

"I try to get them to be in the moment," she says. "They shouldn't be complicating their performance by thinking, 'Fix this, change that'—they've already done their work at that point. I try to get them to focus on one or two things that they relate to very strongly, something like: 'I want to give something to the audience,' or 'I want to enjoy myself,' or 'I want to be in touch with the music.'" The concept is very similar to the "holistic process goal" that Tim Woodman recommends for choke-prone athletes.

Above all, says Hamilton, the point is for performers to enjoy what they're doing. "I don't think anyone can get in trouble by having fun," she says. "After all, they're doing what they love to do—performing."

For Laurence Olivier, the five-year bout with stage fright ultimately yielded not to therapy or to drugs but simply to the passage of time, and the mind's ability to find its way through the tangled web of anxiety. A triumphant performance in the *Merchant of Venice* somehow propelled him clear of his self-doubt. "I had, unbelievably, endured five and a half years of agonizing dread," he wrote. "For those demigods Thespis, Mnester and Genesius to level this particular bolt at my head when I was approaching sixty seemed to me almost too cruel."

But at long last, he wrote, "*it* was over."

Olivier's triumph was not to last long. His stage career was nearly finished. His health had never been strong, and in 1974 he was diagnosed with dermatomyositis, a connective tissue disease. He remained in poor health until his death in 1989, at the age of 82. When it came to stage fright, though, he always remembered that he had been the victor. "The greatest benefit of all from this blessed relief was that now I could feel free to retire from stage-acting," he wrote, "without the personal trauma of knowing for the rest of my life it was fear and not choice that had driven me from my principal métier."

FEAR DOES NOT always make itself known. It can choose to appear via a path so roundabout that it leaves its true origins mysterious. In the case of performance anxiety, such as Olivier experienced, it manifests as an experienced performer's sudden inability to execute a well-learned skill. In the case of panic disorder, it erupts in the form of physiological symptoms so severe that the sufferer often identifies it as a heart attack or some other kind of physical breakdown. In all of these manifestations, the true nature of the fear is so deeply obscured that sufferers are at a loss to seek an effective remedy. Instead, they indulge in superstitions or withdraw from places or things associated with their symptoms. In either case, the underlying fear continues its torments, often for years. That's doubly unfortunate, because anxiety disorders are eminently treatable. The first and most important step in handling the problem is simply to identify the underlying cause. Once the problem is given its correct name, the battle is half won.

Not all fear is so tractable, however. For the rest of the book, we're going to turn our attention to the kind of fear that comes not from within, but from real, external threats—things like mountain lions, angry bosses, and airplanes with broken wings. We're going to delve into the various strategies that people can use to battle fear, and explore how some can backfire. Ultimately, our goal is to understand not just how to hold our own against fear, but to truly master it: to be able to harness its power so that we can reach the highest levels of performance under pressure.

PART THREE

COUNTERATTACK

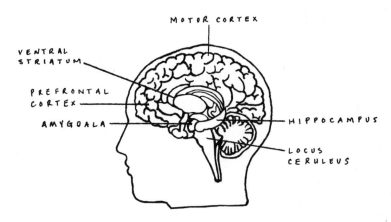

Extreme fear shuts down the prefrontal cortex, where conscious planning takes place, and shifts control to automatic processes generated in such regions as the amygdala and ventral striatum.

Art by Sandra Garcia

CHAPTER NINE

FORCE OF WILL

"NOT BEING ABLE to govern events, I govern myself," wrote the French philosopher Michel de Montaigne. Nearly five hundred years later, his sentiment is echoed in the advice offered by disaster preparedness experts. *Don't panic*, they say. *Stay calm. Keep your wits about you.* This type of strategy is eminently sensible, as there are few situations in which screaming and running around improve one's prospects. But as Montaigne was aware, governing oneself is hard. Is it really possible to master one's emotions? Is panicking or not panicking a decision that's ours to make?

The struggle to master our fears is part of a psychological process called "self-regulation," by which the brain's higher executive functions, centered in the prefrontal cortex, work to restrain the feelings, thoughts, and urges that spontaneously arise in the course of our daily lives. Using willpower to control our mental state is a process that feels a good deal like using our muscles to lift a heavy weight. It takes conscious effort, and as time goes by we grow increasingly tired, until eventually our strength gives out.

The analogy is not perfect. You know ahead of time whether a barbell will be too heavy to lift, just by looking at it. Emotional control is harder to prejudge. Most of us can't say beforehand whether we'll have the willpower to overcome an impulse or a surge of desire. Never having been a best man before, I can't say whether I'll be able to keep from stammering when I stand up to deliver a toast at my brother's wedding—or know, if I fail, whether it was because my willpower was too weak, or because I just didn't try hard enough.

In one important respect, strength and willpower are the same. When the chips are down, we can find that our reserves are far more than we could have imagined.

THE COLDEST, DARKEST, MOST INHOSPITABLE SPOT on the face of the earth is an obscure spot on the featureless whiteness of the Antarctic ice cap, eight hundred miles from the South Pole along the 107th meridian. It was here that the coldest outside air temperature was ever recorded: minus 128.6 degrees Fahrenheit—20 degrees below the freezing point of carbon dioxide. There is no place more inhospitable to human life. Yet this place, Russia's Vostok base, has been inhabited continuously for more than 50 years.

During the late 1950s and early '60s, the Cold War took on new and more literal meaning as the Unites States and Russia expanded their rivalry into the heart of Antarctica. In 1956, a year before Sputnik, the Soviets built their first station, Mirny, on the ice-bound coast. Three days later, the Americans opened their first base, McMurdo Station. The United States followed up later that year with the Amundsen-Scott Station at the South Pole. The Soviets countered with Vostok Station, located 800 miles away at the "pole of cold."

Conditions were brutal for the workers, who had to endure not only unheard-of temperatures and never-ending gales but complete darkness for up to four months at a time. The small crews struggling through the winter in remote stations were more isolated than if they had been in outer space: Once the winter darkness descended, they would be completely isolated for months, with no way for anyone to get in or out.

On April 30, 1961, while a blizzard raged around his tiny base, a twenty-seven-year-old Soviet physician named Leonid Rogozov grimly prepared to perform surgery under what can modestly be described as trying conditions. Abdominal surgery is a difficult procedure in the best of times, with great potential for dangerous complications. And these were far from the best of times. Rogozov's facilities were make-shift and barely adequate; his tools few; his patient in bad shape and deteriorating fast. But all of these problems were minor compared to the biggest obstacle of all: Rogozov's patient was himself.

Rogozov had come to Antarctica as part of the Sixth Soviet expedition, which had arrived three months before to establish a third base, the Novolazarevskaya Research Station. On April 29, with the autumnal nights quickly getting longer and a fierce storm raging all around, Rogozov began to feel ill. He woke up that morning feeling lethargic, and got weaker as the day went on. He became more and more nauseous. A faint pain in the upper part of his abdomen grew and shifted to the lower right part of his belly. He started running a fever. The diagnosis was obvious: He was suffering from acute appendicitis. Left untreated, the infection would weaken and swell the lining of his intestine until it catastrophically ruptured, flooding the internal cavity of his abdomen with feces and pus and swamping his bloodstream with toxins. Death would be inevitable, and agonizing.

Rogozov took medicine to fight the infection, but the next day his condition deteriorated. His fever soared and he vomited repeatedly. The pain was getting worse, and he knew that conditions were ripening for a fatal rupture. "An immediate operation was necessary to save the patient's life," Rogozov later wrote. "The only solution was to operate on myself."

The doctor had no reason to believe that such a feat was possible. If he failed, he would certainly die. To get him through, he would have to rely on nothing more than local anesthesia and willpower fortified by necessity. Gathering the other members of the station, Rogozov laid out a plan. Two atmospheric scientists would be in charge of sterilizing the surgical equipment. A driver would stand next to the doctor during the operation and hold a mirror so that he could see inside his abdominal cavity. A meteorologist would hold the retractors to keep the incision open. Rogozov showed them how to inject him with drugs in case he lost consciousness, and how to administer artificial respiration.

The operation began at ten o'clock at night. Rogozov, dressed in a surgical gown and mask, lay inclined on the operating table, angled to the side so that most of his weight was on his left hip. First he injected a solution of novocaine, a local anesthetic, into his abdominal wall just above the infected appendix. Once the painkiller took effect, he asked one of his assistants for a scalpel and made a five-inch incision. Working by feel, he cut through skin, muscle, and fat tissue, and then used the mirror to watch the knife as he came to the delicate business of exposing the peritoneum, the membrane that surrounds the internal organs. After half an hour, he felt weak and dizzy and had to stop for a rest. Pressing on, he located the appendix, a thumb-like projection from the intestine, and found that it had developed a perforation an inch across. Cutting out the infected organ, he doused the surrounding area with antibiotics and sewed the wound shut. The whole operation had taken a little less than two hours. Rogozov's willpower had pulled him through. He had survived.

In the aftermath, the doctor's condition was quite poor, but as the days went by he improved quickly. After five days his temperature was back to normal and the inflammation in his belly had subsided. A week after the operation he removed the stitches, and a week after that he was back at work. A photograph taken after his recovery shows him seated on a snowbank next to a penguin, smiling and looking hearty.

HUMAN WILL CAN OVERCOME incredible obstacles, as Leonid Rogozov's story demonstrates. But before we can figure out how willpower works, we first have to define it. The concept turns out to be one of those things that's stranger the more closely you look at it. Why, after all, should the mind have to struggle against itself at all? Shouldn't an organism simply be able to do what it wants or needs to do?

The answer lies in the fact that, although we experience the world through what seems to be a unitary consciousness, the brain actually operates as a number of functionally distinct modules, all running simultaneously in parallel. Each has its own purpose and its own goals. These modules may compete with one another, or cooperate, but ultimately they work together to produce a mental experience that seems unified and seamless.

The need for willpower arises from the way that the modules are organized. Matthew Lieberman, a psychologist at the University of California Los Angeles, has proposed a taxonomy of mental activity that makes the issue particularly clear. He divides all mental activity into two types. The first he calls the X-system, after the "x" in "reflex-ive." These automatic processes run quickly and effortlessly outside our conscious awareness. Some are instinctual, like breathing or feeling fear; some are learned, like tying one's shoe. A key feature of X-system processes is that they're efficient and run in parallel, so that we can do several of them at the same time. Some automobile drivers are capa-ble of chewing gum, talking on a cell phone, and changing lanes all at once—and still have enough attention left over to notice the police car lights in their rear-view mirror. Their X-system processes require no conscious effort or exertion of willpower.

Complementing the X-system is the suite of faculties that Lieberman calls the C-system, after the "c" in "reflective." The C-system makes up the part of the brain that engages in reflective consciousness. Its most important region is the prefrontal cortex. Compared to the X-system, the C-system is flexible. Its job is to reflect; to conceptualize; to figure out. It can look into the future and predict the result of behavior that's happening right now. While the X-system runs effortlessly through its repertoire, responding reflexively to the environment, the C-system monitors what it is doing, keeping an eye out for trouble and jump-ing in to supervise when things look like they might be going off the rails.

The C-system is the masterpiece of humanity's evolutionary jour-ney. It's the suite of mental systems that make us sentient beings. We could not write the Constitution, paint the Mona Lisa, or land on the moon without the C-system. But it has its drawbacks. It is slow—positively lethargic, by the standards of the X-system. It takes at least half a second for events in the outside world to make an appearance in the consciousness; anything that happens more quickly—like adjust-ing the swing of a bat to hit a pitch—can only be carried out by the X-system.

The C-system also can only handle a small amount of information at a time. Although it has access to the entirety of declarative memory, it can only process what's stored in working memory—about seven things, give or take. The X-system, in contrast, has access to the subconscious emotional memories stored by the amygdala and can process many pieces of data at once. The X-system is expert at spotting patterns across broad swaths of information. When we stop trying to think of a solution to a puzzle, and then later find the answer popping into our head out of the blue, that's the result of the X-system beavering away out of sight.

Above all, the C-system takes effort. Its processes require the expenditure of mental energy. After we use our C-system vigorously for a while, we feel tired. An hour spent solving difficult math problems, or answering questions in a job interview, or navigating the highway interchanges of New Jersey, leaves us feeling wrung out. Given a choice between doing something the X-system way or the C-system way, most of us will unhesitatingly choose the former, the automatic way—which is another way of saying that we're creatures of habit. Indeed, psychologists speculate the vast majority of the behavior that we engage in during the course of an average day is automatic.

And this brings us back to the subject of willpower. According to Lieberman's model, the exertion of willpower is simply the C-system attempting to forcibly override a strong and deeply ingrained X-system process. Lieberman has identified one brain region in particular—the ventral lateral prefrontal cortex (VLPFC)—as the heart of the C-system, as it plays a key role in many kinds of self-regulation. Brain-scan studies show that the VLPFC becomes active when people suppress the urge to take risks while gambling, when they forgo small immediate rewards in favor of bigger payoffs later, and when they prevent themselves from performing habitual actions. "It seems to be a bit of a one-stop shopping spot for various kinds of inhibitory processes," Lieberman says.

The region's role in controlling fear and other negative emotions was demonstrated in an experiment by Kevin Ochsner, a neuroscientist at Columbia University. Subjects were put in a functional magnetic resonance imaging (fMRI) machine and shown disturbing or frightening images, and either asked to look at them passively or to try to think of them in ways that made them seem less negative. Ochsner found that in the first case there was a notable activation in the amygdala, but when the subjects tried to control their negative emotion, the VLPFC was turned on and the amygdala was deactivated. Indeed, the more strongly the VLPFC was turned on, the less the amygdala was. The C-system, in effect, was shutting down the X-system. Willpower was putting the brakes on fear.

This kind of emotional control takes effort. Roy Baumeister, a psychologist at Florida State University, has hypothesized that self-control is dependent on some kind of resource—a sort of psychic fuel—that gets depleted as one struggles to control one's behavior, thoughts, or emotions. When subjects in an experimental setting are called upon to perform two acts of self-control, their performance on the second one is impaired. For instance, if asked to repeatedly squeeze a handgrip until it becomes too painful to continue, they will give up sooner if they've first been asked to suppress an emotional response, such as in the Ochsner study. "The exact nature of the hypothesized resource is not known," Bauman writes. But it acts and sounds a lot like willpower.

Says Lieberman, "We think that what he's talking about is the right VLPFC."

CONTROLLING OUR EMOTIONS by force of will gets harder as we get tired. It also gets harder as our level of fear increases. Sue Yellowtail experienced this phenomenon when she was stalked by the mountain lion. When the animal was relatively far away, she was able to keep a leash on her impulse to run away. But once the cat got close enough to pounce, Yellowtail could no longer restrain herself. She panicked and ran. As we've seen, under conditions of intense fear, the amygdala activates the locus ceruleus, which releases high levels of noradrenaline in the prefrontal cortex. These elevated noradrenaline levels trigger the α1-receptor, which works to deactivate the whole of the lateral prefrontal cortex. In essence, the fear system pulls the plug on the cognitive processes that normally keep it in check. Lieberman writes: "When threats are imminent and arousal is high, our brain takes the decision out of the hands of the C-system and puts it in the hands of the X-system."

We all have a breaking point, then, beyond which deliberate, conscious control of behavior inevitably gives way to freeze, fight, fright, or flight. Unfortunately, most of us have a hard time appreciating before the fact how nonnegotiable this effect will be. We get so used to making our way through the world under the stewardship of our complex and sophisticated C-system that we tend to assume that we will always have it at our disposal. When we suddenly find ourselves drowning in a flood of noradrenaline, it can be shocking how little brainpower we have at our disposal.

The failure to anticipate the cognitive effects of extreme fear can result in tragic consequences. Many passengers in jet airplanes ignore the preflight safety briefing, figuring that if worse comes to worst they'll be able to figure out the briefing card from the seat pocket in front of them. Such confidence is misplaced. The odds are poor that a crash survivor will have the cognitive wherewithal to figure anything out in a burning, smoke-filled enclosure. Conversely, the National Transportation Safety Board has found that passengers who take the time to read their airline safety card are significantly more likely to avoid injury or death in case of an accident. Having used their C-systems before the crisis erupts to figure out what to do, they'll be able to do what's necessary without trying to rely on logical reasoning amidst chaos and screaming.

It doesn't take a disaster for failures of self-control to turn deadly. It can happen in the course of any activity that calls for self-reliance in a potentially dangerous setting. Take scuba diving. The underwater environment is inherently stressful; add in the strange sensation of weightlessness, the cold of the water, the constriction of the wetsuit and other gear, and the restricted vision, and the psychological load is even heavier. One of the major side effects of stress is rapid breathing. In the grip of panic, a person fighting for air will instinctively rip away anything that's blocking his mouth. Unfortunately for a scuba diver, that thing is the regulator that supplies his air. Thus a situation that in itself might only be mildly stressful or scary can quickly spiral into a fatal spasm of panic. Sports psychologist William Morgan, who spent ten years studying scuba panic at the University of Wisconsin, suggested that panic may play a part in as many as 60 percent of all underwater fatalities.

A particularly dangerous variant is cave diving. As we've seen in the chapter on panic attacks, the sensation that one is trapped inside one's environment can be intensely stressful. This is bad enough when you're under a hundred feet of water, but all the more so when you find yourself in a long, dark cave, separated from the safety of the surface by solid rock. Worse, underwater cave bottoms are often covered in fine silt, which is easily stirred up to produce a disorienting blackout effect, very much like the spatial disorientation effect that can be so deadly for pilots. When you're scuba diving deep in a cave, the slightest twinge of anxiety can quickly spin into a catastrophe from which there is no escape.

ON EASTER SUNDAY, 1992, a twenty-six-year-old Australian spelunker named Rolf Adams prepared to enter the water of Jackson Blue Spring, a long, shallow lake in the Florida panhandle whose crystalline waters conceal an intricate network of caves. An expert in exploring dry caves, Adams had signed up with cave explorer Bill Stone for an expedition to the bottom of Sistema Huautla, a labyrinth of flooded caverns that lies beneath the Sierra Mazateca mountains near Oaxaca, Mexico. It was the last day of a five-week training camp. The next morning, they were scheduled to leave for Mexico. Adams was eager to get as much practice in as possible.

Accompanied by Jim Smith, an experienced cave diver, Adams entered the water and finned to the entrance of Hole in the Wall, a cave whose mouth lay at the bottom of a bowl-like depression at the lake's bottom. As the men descended through the slot-like entrance, the bright sunshine of the surface dimmed to a murky glow. Straight down they plunged into the preternatural stillness, past walls of pock-marked stone garlanded with the disks of long-dead sand dollars. Sixty feet down, the darkness was almost complete. Playing the beams of their flashlights over the cave walls, they located the end of a fixed line that ran all the way to the cave's bottom.

Adams and Smith followed the line in, passing through a series of narrow passageways and domed chambers. Blind white salamanders scurried past, disappearing into the inky darkness. An eel, disturbed by the beam of their lights, wriggled away.

After finning about two thousand feet into the cave, the men turned around to come back out. When they were still a thousand feet from the cave mouth, Smith turned to check on his buddy and found that he had floated to the roof of the cave, where he was exchanging his primary regulator for his backup one. Having accomplished that, Adams signaled to Smith that he was okay. Smith turned and continued to swim for the entrance. Moments later Adams caught up with him, gesturing frantically: He was out of air. Calmly, the experienced Smith handed Adams his regulator to breathe from, and began breathing from his own backup regulator. As the switch was underway the two men, distracted, were unable to maintain their buoyancy, and slowly sank to the cave floor. Stirred by their motion, the fine silt that had gathered there over the years billowed up and surrounded them. In the confusion the men overcompensated for their buoyancy and floated to the cave ceiling. The regulator fell out of Adams' mouth. He pulled away from Smith and disappeared into the murk.

Managing to reorient himself toward the entrance, Smith swam for safety and barely managed to make it clear of the cave mouth with virtually no air remaining in his tank. When Adams' body was recovered, his equipment turned out to be working just fine, and his tank had plenty of air. What had killed him wasn't his lungs, but his brain—inexperience and stress compounded by disorientation, claustrophobia, and sensory deprivation.

Panic is far from rare among scuba divers. One survey conducted by Morgan found that more than half experience panic or near-panic at least once. A cave diver's vulnerability to anxiety and panic is so amplified that even the most steely nerved veteran is in danger of succumbing. Indeed, Morgan found that divers with extensive experience could find themselves panicking for no obvious reason. A notable case in point was the man who was sent in to recover Adams's body, Sheck Exley. Tragically, Exley himself was found dead two years later at the bottom of another deep cave—despite having twenty-nine years' experience and being regarded as one of the most skilled and knowledgeable cave divers in the world.

Probably anyone can die from panic while cave diving. But having studied numerous such cases, Morgan concluded that a certain personality type is particularly at risk. In a 2004 study, he followed the progress of forty-two students as they completed a four-month introductory scuba course, and found that those who scored high on a psychological measure called "trait anxiety"—basically, those who have an anxious personality—were far more likely to suffer a panic attack. Indeed, Morgan found that he could predict with 83 percent accuracy whether a beginner diver would or would not panic in the course of their training.

It may not come as a shock to learn that anxious people are more likely to panic, but what's more intriguing is the recommendation that Morgan had for these divers. He found that no therapy could help them overcome their anxiety enough that they would be safe underwater. If these divers try to calm themselves in the face of panic—if they try to regulate their emotion—they will instead likely trigger a condition called "relaxation-induced anxiety" (RIA). The harder they try to quell their fears, the faster they will panic. As with panic attack sufferers, the mere fact of being aware of their anxiety, and their bodies' reaction to it, is enough to send their sympathetic nervous system spiraling out of control. The best option for divers with trait anxiety, Morgan concluded, is to just stay out of the water.

Relaxation-induced anxiety is a subset of the paradoxical effects that Daniel Wegman has been studying. As you'll recall, paradoxical effects occur when the mind attempts to achieve a goal and winds up achieving the opposite. They're like elderly drivers who think that they're hitting the brake but instead hit the gas and plow through a plate-glass window.

Paradoxical effects might sound bizarre, but they are not rare. Indeed, they are a recurring feature of attempts at self-control. People who diet and gain weight, or lie awake at night counting sheep, are at the pointy end of a paradoxical effect.

Attempts to suppress fear through the exertion of willpower seem to be particularly prone to paradox. Just as trying not to think about a white bear makes the image spring up into our thoughts willy-nilly, trying to avert our attention from something we're scared of just makes the fear pop up somewhere else. It's like a mental Whack-a-Mole.

It's even possible to induce paradoxical stress in other people. German psychologist Sonja Rohrmann led a study in which participants were asked to speak in front of a group of people. Hooked up to equipment to measure their physiological response, the subjects were split into three groups and instructed to wait for a few minutes before they gave their talk. The first was allowed to wait in silence, while the second was informed that their heart rates were high. The third was told that their heart rates were low. This latter group—those falsely assured that they were showing no signs of stress—were not calmed by this (erroneous) news. On the contrary, their heart rate, blood pressure, and cortisol levels were the highest of all three groups. Reassuring them, then, only made them more tense. While the researchers found no definitive explanation for this phenomenon, they surmised that being told that they were calm when they felt otherwise made the subjects feel even more nervous about the disquieting bodily sensations they were experiencing.

Take-home lesson: Trying to tell yourself that everything is fine when your X-system's alarm bells are going off is not only a fruitless strategy, it can actually makes things worse.

WILLPOWER, IT SEEMS, is not very effective at fighting extreme fear. It's too fraught with drawbacks. How, then, can we explain a phenomenon like Rogozov? How could he have used the force of his conscious mind, with all its limitations, to stand fast in the face of pain and danger?

We may never know for sure. Rogozov might simply have been a one-in-a-million, a man who happened to possess an extraordinary gift of willpower and stumbled upon the occasion to need it. But there's another possibility. Rogozov may have relied on more than sheer willpower to get him through his surgery. When it comes to battling fear, there's more than one way to skin a cat. While it may be fruitless to battle the X-system head on, there are ways to work around its weaknesses, take advantage of its blind spots, and harness its strengths. Perhaps Rogozov was able to make use of such a strategy.

We all can aim for the same goal. Understanding how the emotion of fear is generated in the brain, we can develop more sophisticated strategies to sidestep the difficulties that would arise if we merely grappled with it head on. First, we can use these insights to prepare ourselves. Then, when we're in the grip of fear itself, we can make the best use of our mental resources. By being smart, we can avoid having to be strong.

CHAPTER TEN

STEELING YOURSELF

IT'S IMPOSSIBLE TO see more than a few feet through the thick smoke, but what I can make out is bad enough. The deck has buckled upward, and fragments of tables and chairs lie piled among broken steel beams. Flashes of light illuminate a severed half body dangling from a pyramid of debris that reaches to the ceiling. Trapped and wounded men are calling out from all around, their moans punctuated by the groan of twisting, ruptured metal. I'm on my knees, doubled over, struggling with the weight of a stretcher as I maneuver under a section of fallen decking. Our ship his just taken a direct hit from an enemy missile, and my teammates and I have to extricate as many casualties as we can from the impact zone.

The wounded man doesn't sound good. He's gasping for breath, trying to get oxygen into his smoke-damaged lungs. "Euh-uh-oh!" I'm finding it hard to breathe myself, through the sweaty flash-protection hood that covers my nose and mouth, and I'm struggling to see through my helmet's fogged-up Plexiglas face shield. Hopefully, my team will soon find an exit hatch—one without a raging fire behind it. We've got to get this guy out of here without killing him, or getting ourselves killed in the process.

We find ourselves in a debris-free corridor and stop for a head count, our arms burning from the weight of the injured man. Four stretcher bearers and two other men make six. Where are our other three team members? At this moment our petty officer emerges out of the smoke. "What?" he erupts. "You let your team get separated? I don't believe it! Wait here!" He storms off to find the missing men. We all stare at the floor, exhausted. Letting teammates get separated in the zero-visibility conditions of a shipboard fire is as good as signing their death warrants. We've screwed up, big-time.

We're in trouble, but not in actual danger. My teammates and I are struggling through a twelve-hour shift aboard the USS *Trayer*, a full-scale replica of an Arleigh Burke–class guided-missile destroyer. The $83 million vessel rises above a 90,000-gallon tank of water inside a building at the Navy's Recruit Training Command facility, 28 miles north of Chicago. Opened in 2007 after ten years of planning and construction, it's the most realistic and ambitious naval simulator in the world.

Officially commissioned as a full-fledged vessel, the *Trayer* has jokingly been dubbed "the unluckiest ship in the Navy." Five days a week, it bears the brunt of multiple enemy missile attacks that result in deaths, injuries, and potential sinking. But while the scenario is make-believe, the stakes are real. For recruits coming off their eight-week basic training course, the *Trayer* is their baptism of fire and their final exam. It's all that stands between them and the right to be called "sailor."

The Navy has allowed me to take part in a simulator run as an embedded journalist, to see what their newest training facility is like from the inside. I arrive at the base a few minutes before the scheduled 8:45 p.m. start time, hurriedly change out of my civvies into a blue Navy jumpsuit, and fall in with a phalanx of recruits, 250 strong, standing outside a hangar-like building. An officer delivers a speech about honor, courage, and commitment. The recruits roar "Booyah!," the doors swing open, and we march inside.

Suddenly we're on a pier, the *Trayer* rising above us. Though we're indoors, a gentle breeze is blowing that carries the distinctive tang of the sea. Seagulls caw somewhere overhead, and waves lap between the pier and the hull of the ship. We climb up the gangway in teams, then descend into the bowels of the ship.

The night starts slowly, with exercises that involve monitoring gauges, standing lookout, and coiling rope. Every 45 minutes or so, we return to our "crew living space" to visit the head and to watch videos describing U.S. Navy history. Then, late in the night, just as the petty officers start chewing out recruits for drifting off...BOOM!

Giant subwoofers buried in the floor set us shaking as the lights briefly go dark. "Brace for shock!" a petty officer shouts. We dash to the sides of the room and lean against the walls, then race to don our battle dress: fire-resistant hood, helmet with face guard, long gloves. The loudspeaker announces "General Quarters." For the next four hours we'll dash from one hair-raising scenario to another, including escaping from a smoke-filled compartment through an emergency hatch and hauling 35-pound canisters of 5-inch gun ammunition out of a room slowly filling with water.

Sometime around dawn, as the ongoing battle continues to rage, we strap on oxygen bottles and don masks and heavier hoods to fight a fire inside one of the damaged compartments. Flames roar in a propane-fed inferno as the team holds a heavy pressurized fire hose. The man at the front plays the nozzle back and forth over the fire. From time to time a hidden speaker blares the sound of a small explosion, representing a barrel cooking off from the heat, and we're doused with a spray of water.

As the night continues, the catastrophes grow more taxing. Young and inexperienced, the recruits struggle to make the most of their two months of naval education. Our screw-up in the casualty exercise is the low point. After we finish lugging the groaning dummy up to the sick bay the instructors tear into us. "Look, he's dead," a petty officer says as he rips open the stretcher's Velcro straps to show us how we fastened them incorrectly. "You killed him."

The death of our wounded comrade is so demoralizing, the depths of our fatigue so dispiriting, that we're hardly consoled by the fact that our dead shipmate is made of plastic. But that's the point. Though the danger might be simulated, the emotions we're feeling are real.

The *Trayer* is just one part of a massive and ongoing effort by the U.S. military to provide highly realistic training simulators for every branch of service, from life-sized desert towns where ground troops can practice urban combat, to electronic simulators that realistically mimic the handling characteristics of supersonic jet fighters and ballistic missile submarines. In the armed forces, more than any other sphere of modern life, performance under extreme stress is crucial. It's more than a matter of life and death; it's a matter of victory versus defeat. It's incumbent, then, on the military leadership to prepare every soldier, sailor, and airman not only for the specific duties that they'll be trained to carry out, but the secrets of functioning in the strange parallel world of extreme fear. The first time they "go to see the elephant" shouldn't feel like the first.

Those of us who aren't serving in the armed forces might not face life-or-death crises as often as combat personnel, but we still have to deal with fear, and we have the same set of mental equipment between our ears. So for all of us the issue in facing fear is the same: How can we maximize our performance under stress and keep our fears from running away with us?

In this chapter and the next, we'll explore some of the techniques that have been developed for keeping it together in the face of extreme fear. Some of these strategies go back thousands of years, while others have been elucidated only recently. Together, they can serve as a mental armor against panic and against fear in general.

HABITUATE

The most powerful tool for handling fear is so basic that virtually every animal makes use of it in some form. The psychological term is "habituation," but the idea is simple. You get used to things. Explosions and wreckage are shocking the first time you experience them, but spend enough time in a simulator like the *Trayer* and they will eventually become old hat.

The ability to habituate is essential to the functioning of any nervous system. Animals move through the world looking for good things and avoiding bad things. Everything else has to be ignored, or else their neural circuits will be flooded with useless information. So animals are attuned to novelty. New things get looked at, sniffed, poked, and tasted. Things that are familiar eventually get ignored. Even the sea slug, which has a primitive nervous system but no brain, can habituate. A poke to its feathery gills and feeding siphon normally causes them to retract instantly, but if they are poked long enough, the sea slug learns to leave its appendages fearlessly dangling.

Human beings, of course, can habituate to a much wider variety of sensations—for better or worse. We learn to tolerate unpleasant things, but also lose the intense rapture of novel pleasures. The first time you taste expensive chocolates, your taste buds swoon, but if you eat them every day they become routine. You love wearing a new shirt, but each time you wear it the joy gets a little less. You like your music really loud, so you crank it up until your ears hurt, then habituate to the noise level, and crank it up some more. Eventually the foundations are shaking and your neighbors are calling the cops, but to you it feels just as loud as when you started.

We habituate to fear somewhat differently, through a process called "extinction." The amygdala's memory system retains frightening associations permanently. If you've been bitten by a white dog, your amygdala will never forget. But the memory can get overlaid by a positive or neutral association. If you later buy a friendly white dog, say, and spend day after day associating your pet with harmless fun, in time your fear of white dogs will be overlain by a suppressing response generated by the medial prefrontal cortex. Your amygdala isn't forgetting that white dogs are dangerous, but you've layered a new memory on top of it, like a linoleum floor over a trap door. The old unconscious fear connection remains, encoded subconsciously in the amygdala. Under stress, the buried fear can spontaneously reemerge.

Despite this drawback, habituation is one of the most effective tools we have for preparing ourselves for fear. When you expose yourself to whatever you're afraid of and no harm ensues, your prefrontal cortex and hippocampus are able to overwrite the amygdala's fear association. In essence, your C-system modifies the X-system response. Habituation has long served as the bedrock of military training. Sixty years before the USS *Trayer*'s hi-tech simulated noise, smoke, and gore, the Army put recruits into a "mental conditioning chamber" in which they listened to recordings of screams and twisting metal while sand and water was blown in their faces and the smell of a decomposing horse carcass was wafted over them.

The tradition goes back much further than that, probably to the very beginning of military training. "It is immensely important that no soldier, whatever his rank, should wait for war to expose him to those aspects of active service that amaze and confuse him when he first comes across them," wrote Carl von Clausewitz, the great nineteenth-century military theorist. "If he has met them even once before, they will begin to be familiar to him."

Merely exposing a person to an object of fear is not sufficient for proper habituation, however. The most effective way is to experience the stimulus frequently and steadily. As we've seen, the X-system is a slow learner, so habituating to strong fear can take a long time. The key is repetition. Londoners who were subjected to German bombings regularly during the Blitz eventually grew blasé. They habituated to the wail of the air-raid sirens, the ritual tramping down into the bomb shelter, the rumbling thuds of distant explosions. The terror of aerial devastation, which prewar theorists had predicted would quickly cow a populace into submission, instead became a commonplace, a part of daily life. Conversely, infrequent or irregular exposure to fear may not lead to habituation at all, but to its opposite: sensitization. Instead of growing smaller, the response to a stimulus grows more intense. Britons who lived in the suburbs, and who thus were only sporadically exposed to the dangers of war, were much more terrified of German raids.

The same effect helps reinforce anxieties and phobias. People who fear spiders, for instance, may understandably take pains to avoid them, so that on those rare occasions when they are forced to confront one, their hearts race and their guts churn with all the unpleasant symptoms of a panic attack. Thus their fear is bolstered: Spiders really *are* scary. To get around this reinforcement effect, therapists treat phobias by regularly exposing their patients to their fears. First, an arachnophobe might be asked to think about a spider, then to look at a picture of a spider, and then to approach a spider inside a glass cage. At each stage, the patient is able to habituate to that level of intensity, before ratcheting up the intensity of the exposure, so that at no point is the stress intense enough to trigger a panic and reinforce the phobia. This kind of therapy has one of the highest success rates in all of psychotherapy, with close to 100 percent of all patients substantially cured.

Habituating to fear can have a dark side. Psychologist Thomas Joiner of Florida State University argues that habituation is a key step that people have to accomplish before they can commit suicide. The fear of death is so powerful and innate that it is nearly impossible to overcome on one's first try. That's why most people who successfully kill themselves are veterans of several previous attempts; after each attempt, the act becomes a little bit easier, and they're able to go a little further toward their goal. Some people, Joiner points out, don't need as much habituation because they are already inured to violence and bloodshed in the course of their professional lives. And that, Joiner says, is why cops and firemen have suicide rates far above average.

TRAIN

When we're faced with extreme danger, our C-system shuts down and our automatic X-system takes over. This can be disastrous if the X-system is incapable of dealing with the situation. But an X-system takeover might be just the ticket if its automatic processes are appropriate for the challenge at hand. For the men and women beginning their careers as sailors aboard the USS *Trayer,* the ability to perform in the heat of battle will be achieved through endless rote drill, as it has been for military personnel around the world and throughout the ages. For combat personnel, it's best to assume that the C-system is going to collapse, and therefore to train the X-system so that it knows what to do. The process of performing a required skill over and over until it has become deeply automatized is called "overlearning," and it produces a set of skills that are resistant to all but the blindest panic. As Vegetius, a Roman writer of the fifth century AD, wrote: "Few men are born brave; many become so through training and force of discipline."

Vegetius was writing at a time when the Western Roman Empire was on the brink of collapse. His book *De Re Militari* was an attempt to compile and organize the lessons from Rome's military heyday, in the hope that their application might return his country to greatness. Vegetius believed that the essence of that greatness was the training to which his ancestors had tirelessly applied themselves. "The Romans owed the conquest of the world to no other cause than continual military training," he wrote. "They thoroughly understood the importance of hardening themselves by continual practice. A handful of men, inured to war, proceed to certain victory, while on the contrary numerous armies of raw and undisciplined troops are but multitudes of men dragged to slaughter."

His common-sense approach made Vegetius a favorite of military leaders for more than a millennium, and his principles are still widely applied today. "The more you sweat in training the less you bleed in war," says Rob Smithee, a Green Beret officer who formerly oversaw a Special Operations training program. "The whole reason they get you to march in step, the whole reason they yell at you and get you all charged up, screaming and hooting and hollering, is so when the whistle's blown, and you're told to go over the top and charge into the machine gun nest, there's an automatic response. There's no thinking."

If military trainers have agreed for two thousand years that automatizing a soldier's skills is the surest way to keep him useful under intense stress, a corollary question has been much more hotly debated: Is it better to train to high proficiency before trying to use those skills under stress, or to train from the beginning under intense stress similar to that in which the skills will have to be used? Better, in other words, to drill frightened or calm?

Several studies into this question all support the conclusion that trainees should be allowed to gain proficiency *before* stress is added to the learning environment. Getting a student's cortisol flowing before a task has been fully automatized yields poor results, since high stress shuts down the C-system before the skill has been transferred to the X-system. A study of skydiving students, for instance, found that

novice jumpers had a hard time memorizing lists of words while falling through the air. In their paper, the researchers quoted the wisdom
of one of their skydiving instructors: "No matter how smart you are
on the ground," he said, "you get stupid the first time you fall out of
a plane."

"Motor movement, strategies, tactics, and all aspects of performance first need to be learned under low-stress conditions," concurs
Marc Taylor, a psychologist at the Naval Health Research Center in
San Diego. "That means going from being a total novice to an inter-
mediate-level performer. A Navy Seal has to learn how to tie a knot
under nonstressful conditions before he can know how to tie it 40
feet underwater in 59-degree water. Before the kicker for the football
team can make that game-winning, high-stress 45-yard field goal, he
first has to learn it in practice, when no one's watching."

Once the Seal or the kicker has learned those skills, however,
performing them under realistic stress can provide a double boost,
enlisting the benefits of both training and habituation. Indeed, Sian
Beilock, a psychologist at the University of Chicago who has extensively studied the mechanisms of choking, found that golfers who
practiced putting while being watched by an audience performed
much better than when put into a high-stress putting contest in which
a large monetary prize was at stake. In effect, they had learned not
only to hone their putting moves, but also to become accustomed to
the stress of competition.

Becoming skilled at what one does leads to confidence, itself a
powerful shield against fear. Psychologists use the term "self-efficacy"
to describe the mental state of a person who feels empowered and capable of taking on whatever challenges await. Studies have demonstrated
that this feeling of confidence in one's abilities leads to greater success
amid crisis, even apart from the direct benefits of the skills themselves.
"The courage of a soldier is heightened by his knowledge of his profession," Vegetius wrote, "and he only wants an opportunity to execute
what he is convinced he has been perfectly taught."

Success breeds confidence and dispels fear. Earlier, we came across
a hormonal effect called the "winner's effect," in which those who

triumph in a contest experience a surge of testosterone, which predisposes them to greater aggressiveness and confidence. Winning begets winning. For this reason, Vegetius recommended that military commanders take advantage of this effect to shore up the fighting spirit of demoralized or untested troops.

"There are without doubt some of a more timorous disposition who are disordered by the very sight of the enemy," he wrote. "To diminish these apprehensions before you venture on action, draw up your army frequently in order of battle in some safe situation, so that your men may be accustomed to the sight and appearance of the enemy. When opportunity offers, they should be sent to fall upon them and endeavor to put them to flight or kill some of their men. Thus they will become acquainted with their customs, arms, and horses. And the objects with which we are once familiarized are no longer capable of inspiring us with terror."

STEEL YOURSELF

Training and habituation are both demonstrably effective, but they share a major weakness: they're specific. We habituate to a particular fear and train to perform a particular skill. In the real world, we never know what kind of trouble we might wind up in. Fortunately, studies have shown that we can train ourselves to be more resistant to all stress. We can inure ourselves against fear in general.

First step: exercise. In a pioneering study into the nature of courage, Canadian psychologist Stanley Rachman interviewed dozens of British Army bomb disposal operators, members of an elite unit that for more than a decade waged a deadly cat-and-mouse game with the bomb-makers of the Irish Republican Army (IRA) in Northern Ireland. Their work required skill and patience in the face of tremendous danger; in the course of defusing more than 30,000 explosive devices, 17 operators were killed. Curious to find out what separated the most courageous operators from their peers, Rachman studied various psychological tests that had been administered to the men in the course of their service. "To our surprise, we came across a feature that distinguished the operators who had received decorations for gallantry," he wrote. "The decorated operators were found to be slightly but significantly superior in all-round psychological health and bodily fitness."

That the difference in fitness was slight may have been due to the fact that, compared to the general population, the bomb disposal operators were all in very good shape to begin with. Even stronger evidence that physical fitness promotes resistance to fear comes from a 2008 study by Lilly Mujica-Parodi and Marc Taylor, in which they further analyzed the hormonal and cognitive response of first-time skydivers, including me. They found that the lower the percentage of body fat, the less stressed out the jumpers were, as measured by the concentration of cortisol in their blood. They also performed better on cognitive puzzle tests that they took just before the jump. "Fitness produces not only physiological stress resilience," says Mujica-Parodi, "but also cognitive resilience. So those individuals actually were able to preserve their cognitive function in the face of a stressor." Which is to say, we all get stupid when we jump out of a plane, but some of us get stupider than others.

Part of the benefit of getting into physical shape may derive from the reinforcement that exercise gives to willpower. As Muraven and Baumeister have demonstrated, the brain's ability to self-regulate behaves in many ways like a muscle, and like muscle, it can be built up over time. If their model is correct, than any forceful attempt to exert C-system control over the X-system will help to build up the brain's reserves of self-control. As most of us can probably attest, getting oneself to exercise on a regular basis requires a lot of willpower. Forcing yourself out the door for a run on a drizzly morning, then, might serve to bulk up two sets of muscles: the ones in the your legs and the ones in your right VLPFC.

Exercise isn't the only way to pump up your self-regulatory powers. It's possible to gain "a sense of self-efficacy by successfully exerting self-control over posture, diet, or mood," Muraven and Baumeister write. "People should improve in self-control ability even after failing at the self-control task."

A general program of courage-enhancement, then, might look a lot like taking care of your general health. You'd want to exercise frequently, exert cognitive control over what you eat, and in general ride herd on your more impulsive tendencies. If you were really gung ho, you could eat raw eggs, shower under cold waterfalls, or pull a rock-filled sledge with your teeth. Totally optional.

Most importantly, if you blew your diet once in a while, or slept late a morning here or there, you wouldn't give yourself a hard time, because according to Muraven and Baumeister, you'd be building willpower even when you tried and failed. The important thing would be not to stress out about it. All kinds of stress are cumulative, whether physical or emotional, and it's their total that determines how well you'll perform. If you're rested and well-nourished, and have generally been leading a balanced existence, you're going to have a greater capacity to handle fear when things get dicey.

Conversely, if you're fighting with your wife, avoiding bill collectors, and working 20-hour days, your cortisol levels won't have far to go before the hormone floods your bloodstream. Remember the case of Cindy Jacobs, who suffered a life-changing panic attack after overextending her family responsibilities, and Laurence Olivier, whose 40-year stage career nearly ended when a combination of personal and business crises triggered his first run-in with stage fright. As much as possible, keep your day-to-day life free of stress and anxiety, and ramp up your sympathetic nervous system only during training and exercise.

Lowering your stress level can be as simple as wearing a warm coat on a cold day. Physical stressors like noise, hunger, heat, and cold can take a major toll in high-intensity situations. William Morgan, the University of Wisconsin psychologist who made a specialty of studying scuba accidents, found that divers who swam in cold water without an adequate wetsuit were significantly more likely to panic.

So to steel yourself for danger, take the advice of an old song: Button up your overcoat and get to bed by three.

PREPARE TO LOSE YOUR MIND

You've done everything you can to brace yourself for the onslaught of fear. You've trained, you've habituated, you've cut your body fat down to 1 percent, and committed yourself to going to bed early every night in an extra snuggly sleep set. Now chaos and disaster are looming, and you're rested and ready. All systems go. There's just one more thing to do: Think.

No matter how you've buttressed and fortified the C-system, there will inevitably be a point beyond which the fear becomes too intense and the amygdala shuts down the frontal cortex. You will become stupid. So if there's anything you'll need to think of in the crisis ahead, the time to think about it is now, before you lose your powers of planning, thinking, and reason.

It's hard to overstate just how differently the mind works when suddenly placed under the stress of extreme fear. I learned this firsthand when I was taking flying lessons at a glider club in upstate New York. I had climbed into the front seat of a Schweitzer 2–33 glider with my instructor seated directly behind me. The tow plane, a Piper Pawnee crop duster, had swung around in front of us and a volunteer had hooked up the tow rope between us. When I was ready, I would give a signal to the volunteer, who would signal the Pawnee pilot, who would then take off, pulling us behind him. Once we reached an altitude of 3,000 feet, I'd pull the tow-release knob, and I'd be gliding free.

Unbeknownst to me, however, my instructor had made prior arrangements with the Pawnee pilot to do a maneuver called a "low rope break." This was a simulated emergency exercise intended to replicate a particularly dangerous happenstance in the sport of gliding: the loss of power on the part of the tow plane at very low altitude. Even the training maneuver is potentially hazardous, because there is little time for the pilot to react and a very small margin of error. And, of course, for the exercise to have any value, it had to be a surprise for me, the student pilot.

The tow rope was hooked up; we were ready to go. I gave the signal, and the Pawnee accelerated. We followed behind, gaining speed, and soon were soaring along five feet over the grass. And then, unexpectedly, the glider started to gain ground on the Pawnee. The tow plane was slowing down. Was it engine trouble? I had the presence of mind to steer to the side and pull the tow-rope release. To my surprise, the glider started to climb. Even though we had no power to pull us through the air, we were rising higher and higher above the airfield— the end of which was rapidly approaching. If I didn't do something fast, a collision into a line of trees would be inevitable.

The solution was obvious, once I had time to think about it later. By my left hand was a handle that I could pull to deploy the speed brakes, a pair of metal slats that spring up on either wing to create drag. There I was, halfway down the runway, too high and too fast and running out of room. Obviously, I needed to deploy the speed brakes. But it didn't even occur to me. I just sat there, dumbfounded, until my instructor yelled "Speed brake!" and, for added measure, pulled the lever himself. Within seconds, we were safely back on the ground.

When we're under high stress, our working memory shrinks and our cognitive focus narrows, and the only way we can make effective decisions is through simple rules of thumb that psychologists call heuristics. A good example is "Don't shoot until you see the whites of their eyes!" It's the perfect kind of instruction to give an untested colonial militiaman who's about to face a column of British Army veterans marching up Bunker Hill. No complicated cognition or subtle judgment is required. Just a simple yes-no, binary call: If you can't see the whites of their eyes, don't shoot. If you can see the whites of their eyes, do shoot. And the rest is history.

During that first unsuccessful rope break, I had no heuristic. But when I went back up with my instructor and tried again, I was armed: "When running out of room over the runway, pull the speed brake." Once again my instructor pulled the tow release, and once again the glider ballooned. This time I didn't need to think. I pulled the speed brake. Piece of cake.

Some situations are too complicated to be handled by a heuristic, though—like operating a complicated piece of equipment. In such cases a more powerful mental prop is required.

In the mid-1930s Boeing developed a new state-of-the-art bomber, the Model 299, for the U.S. Army Air Corps. The aircraft was a technological marvel, able to carry five times as many bombs twice as far as its predecessors. But it turned out to be so complex that it was nearly impossible for humans to fly. Many steps had to be carried out in the proper sequence or catastrophe could ensue. This was difficult enough under the best of circumstances, but in a high-pressure situation it was well-nigh impossible. On its very first demonstration flight before Army brass, the Model 299 roared down the runway, climbed 300 feet into the air, stalled, and crashed. It turned out that while busy with controls for the airplane's four engines, retractable landing gear, flaps, and variable-pitch propellers, the pilot had forgotten to release a lock on the plane's elevator and rudder controls. A newspaper at the time concluded that the Model 299 was "too much airplane for one man to fly."

The Army bought a few of the planes for testing, and a few of the pilots in the program put their heads together to figure out a way to make the plane safe to fly. What they came up with was an idea as simple as the problem was complex. They called it a "checklist." Instead of trying to perform every task from memory, a flight crew would have a written set of instructions that they would refer to religiously. With this mental prop in hand, a crew could perform hundreds of complicated steps even amid the stress of a flak-peppered bombing run. And it worked. The Army wound up buying 13,000 of the redesignated B-17, which became the backbone of the air war against Nazi Germany. And the checklist went on to become an indispensable tool on every aircraft in the sky, from the simplest trainer to the most advanced jet.

Checklists have spread far beyond aviation. Critical-care physicians use them for managing complex medical procedures and scuba divers use them to avoid getting the bends. Many homeowners keep a checklist of sorts beside their telephone: a list of emergency numbers.

Being realistic about how our minds will work while in the grip of fear can mean accepting that we won't be as calm and cool as we'd like, and using that recognition to plan a workaround of our shortcomings. We might even be able to use the fear to our benefit.

In the winter of 1781, a small American force under the command of Brigadier General Daniel Morgan was on the run in South Carolina. An elite British detachment called the British Legion had been sent to wipe them out, in preparation for a combined assault that would destroy the last of the colonial forces in the South and effectively end the revolution.

Morgan knew that his men were inferior to the hardened veterans of the British force. A majority of them were militiamen—farm boys, for the most part—who'd rallied to the cause, bringing with them hunting rifles but little or no military training. In the past such troops had shown a marked propensity to turn tail and flee, sometimes before the battle had actually been joined. But rather than despair at their uselessness, Morgan decided to build a battle plan around their lack of self-control.

With the lead elements of the British force just a few miles away, he called his officers together and laid out his strategy. The place he'd chosen to make his stand was called the Cowpen, a grassy meadow with a hill at one end and backed by a river. Morgan's plan was to station the core of his force, hardened Continental Army soldiers, on the hill. These were the only men he could count on to stand their ground in the face of a British assault. In front of them he placed two lines of militiamen. The first line, furthest out in front, was made up of sharpshooters, outdoorsmen who had honed their marksmanship in the wilds of the New World. Morgan asked them to fire twice at the British, trying especially to hit their officers, and then to retreat behind the hill. The second line he asked to fire three times, and then to retreat and join the first line. The Continentals would then remain to take on what remained of the British force. While the main battle raged, Morgan would then send out a cavalry force that he intended to hide behind the hill, and attack the British from the flank. The militia, too, would come out from their hiding place and rejoin the fight.

By asking his militiamen to fire only a few shots each into the British onslaught, Morgan was recognizing that their self-control in the face of danger, while not nonexistent, was limited. Better to let them retreat in an orderly fashion, and rejoin the fight later, than lose them in a blind panic that could spread to the whole army. Just in case, the river that lay behind the battlefield would prevent them from running too far.

By the time the British arrived at the field the following morning, the colonials had already drawn themselves up into their battle formation. As the British regulars marched forward, the skirmishers unleashed a pair of volleys. Fifteen of the attackers fell. The British line wavered, then regrouped and pressed on as the American line abandoned its position. Beyond lay the second line. As the British attacked, the second line fired its volleys and then it, too, retreated.

The great potential weakness of Morgan's plan was that the make-believe rout might turn into a real one. And for a moment this seemed likely to happen, as the British soldiers began cheering what seemed to be the crumbling of the American resistance. Back in the third line, the seasoned Continentals, not being able to clearly see what was going on in the front, heard their enemy's huzzahs and concluded that the battle was lost. They turned and began retreating up the hill, until Morgan and the Continentals' commander, Lieutenant-Colonel John Eager Howard, were able to regain control, order an about-face, and lead them back down against the British.

The attackers, who had believed the battle won and the rabble fled, now found themselves facing an uphill attack against determined resistance. To make matters worse, Morgan now ordered a double-envelopment attack by his reserves, with a cavalry force sweeping in from the left and the regrouped militia advancing from the right. Caught in the pincer, the British panicked. Many simply collapsed to the ground in shock. The British commander barely managed to escape on horseback, along with a small minority of his men. The British Legion, the cream of the king's forces in North America, had been destroyed as a fighting force. The war's endgame had been set in motion. That summer it would reach its conclusion at Yorktown.

TWO HUNDRED TWENTY-SEVEN YEARS LATER, the Americans aboard the USS *Trayer* are bringing their own battle to a close. Sometime around dawn the captain comes back on the intercom: The enemy ship that attacked us has been captured, and the damage to our ship has been contained. We stand down from general quarters. Amid the relief, we hardly feel our fatigue. It's all downhill from here.

After the recruits march back off the *Trayer,* I'm treated to an extra tour of the ship that real recruits don't get. Scott Barnes, a civilian contractor who oversees the backstage operations, shows me the stage-craft that makes the experience so realistic. From a central control station that looks like the deck of the *Enterprise,* operators can control every aspect of the simulation. In the mass-casualty section, "smoke machines" pump out an aerosol of glycerol that looks just like the real thing, without causing lung damage. Strobe lights mimic the flickering of electrical fires. Heating pads buried inside a steel door mimic the high temperatures caused by a raging fire on the far side. Scent machines electrically heat essential oils to simulate the smell of burning wiring. Speakers hidden inside the dummy's chests relay recorded moans of agony. Several of the companies who helped create the simulator have worked with Disney. Says Barnes, "The *Trayer's* DNA comes from theme parks."

The *Trayer's* intense environment is intended less as a training environment than as an opportunity for sailors to habituate to the sights and sounds of modern sea battle, so that if they encounter a real catastrophe in the future they'll know they can handle it. A major inspiration for the creation of the *Trayer* was the battle to save the USS *Cole* in October of 2000. The Arleigh Burke–class destroyer had docked pierside in the port of Aden, Yemen, and was taking on fuel when a group of Al Qaeda men in a rubber raft drove up to the ship and triggered a massive explosion. The blast blew a 40-by-60 foot hole in the hull, killed 17 sailors, and wounded 39 more. There were no other Navy ships or other U.S. military forces on hand to lend assistance; for hours, the crew were on their own as they fought inside the stricken ship to stop the inrushing water and put out raging fires. The ship was very nearly lost, and the crew struggled for three days before its condition was stabilized. "You can't even imagine the conditions they're living in, and yet they are still fighting 24 hours a day to save their ship and free the bodies of those still trapped," wrote one sailor who witnessed the struggle at firsthand, in an e-mail to a friend back home. "The very fact that these people are still functioning is beyond my comprehension."

As gruesome as the carnage was, it wasn't the first time that the crew had encountered such a situation—or at least, something like it. "The sailors aboard the *Cole* had been through an early-generation battle-damage simulator," says Dr. Mike Belanger, the unit's senior psychologist. "They say that's what saved the ship."

CHAPTER ELEVEN

HANGING ON

IF YOU SUDDENLY found yourself in a life-or-death crisis and had to make a decision that would either save your life or end it, are you confident that you would make the right one?

That was not a rhetorical question for people in the state of Victoria, Australia, during February and March of 2009. For five weeks catastrophic brush fires swept across the state amid record-breaking temperatures and drought. Government policy held that when fire threatened a neighborhood, homeowners were to make a choice: Either stay and fight to save their houses, or evacuate early. They were explicitly instructed not to wait until the flames were close. Trying to run from an advancing wildfire is the surest way to die in it.

The choice given to the people made sense in strictly rational terms. But in the wake of the devastation, a vociferous debate arose over the wisdom of the policy: can people be expected to make rational decisions, critics asked, when they're surrounded by 1,200-degree flames raging four stories high?

Most people have never faced imminent, lethal danger, and so can't possibly know how they will react to the experience of extreme fear. Ensconced within the society's protective cocoon, they've spent their lives free from the need to worry about such things. But as thousands of Australians found out the hard way, danger can overtake the unsuspecting with surprising speed.

Once mortal peril is at hand, the time for mental preparation is past. It's too late to ready one's mind for danger by habituating, or training, or any of the other techniques that we've reviewed. But all is not lost. Although our ability to use reason and logic might be badly impaired, there are still simple strategies available to us that can help stave off panic.

EVERYONE IN Melbourne knew that Saturday, February 7 was going to be brutal. The southern summer had been a scorcher, with temperatures the previous week climbing above 110 degrees Fahrenheit three days in a row. That day the mercury was forecast to climb even higher. Winds were strong and a long drought had left the vegetation brittle and dry, setting the stage for the most dangerous fire conditions ever recorded.

In Glenburn, a farming community outside the city, Victoria University professor Ian Thomas and his wife Bronwen spent that Saturday listening for weather updates on the radio. As an engineer, Thomas specialized in calculating the risk of fire in buildings, so he had a healthy appreciation for the dangers of wildfire. The couple's house and lawn were surrounded by trees on all sides and abutted the eucalyptus forest of Kinglake National Park, which stretched uphill from his front yard. Each summer, fire became a real and present danger.

On this day in particular Thomas was careful to check that the sprinkler system on his roof was in good working order, and that the casks of water that he had positioned around the property were full. "We didn't need the forecast to tell us that it was dangerous," he says, "because when you walked around on the grass, it was crisp under your feet, and when you walked in the bush the leaves crackled. It was obvious that everything was extremely dry."

As it turned out, the weather report was off. The maximum temperature that afternoon in Melbourne wasn't the predicted 111 degrees Fahrenheit, but 115—the city's highest temperature in more than 150 years of record-keeping. The government put out a "Total Fire Ban" alert, which forbade not only the lighting of fires, but also the use of any mechanical equipment, such as angle grinders, that could cause sparks. But all the precautions in the world would offer scant protection given that a single spark ignited anywhere in hundreds of square miles of bone-dry bush would be enough to set off a catastrophe. In any event, it didn't take long. Around 11 o'clock that morning, high winds knocked down a power line that ran through pasture 25 miles to the northwest of Glenburn. Within hours, a roaring wall of flames was burning eastward.

Thomas wasn't concerned over the reports of the fire that he heard over the radio. Given past experience, the outbreak was too far away to pose a danger. A few years before, a fire had started in the national park forest behind the Thomas' house and had burned for a week without getting closer than half a mile.

Around 4 p.m., the scorching heat wave suddenly broke as the fierce, dry northern wind swung around 180 degrees and became a cooler sea breeze. Within minutes the mercury dropped 30 degrees, to a relatively balmy 86. "We started to relax," Thomas says, "because we thought that things were looking pretty good. Nothing big had happened, and that it was likely that there wasn't going to be a major problem." Soon after, the power went out. Fifteen minutes later it came back on, then died again.

What the radio news broadcasts had failed to report was that the wildfire had spread all the way to the town of Kinglake, less than ten miles from the Thomas' house. The new, cool breeze had fanned the flames to new intensity, and was driving the fire toward Glenburn at freight-train speeds.

The first inkling of trouble came when a couple who lived nearby, Lou and Cheryl Newstead, pulled into the driveway. They brought news that their son had just called to tell them that the fire was heading their way. As the couples talked, the wind that was blowing in from the south darkened with smoke. Ash and glowing embers started dropping out of the air.

"We went from not having any particular worries to having fire in our immediate vicinity very quickly," Thomas recalls. The decision point—stay or go—had arrived faster than anyone had anticipated. The neighbors decided to evacuate; the Thomases, to stay and defend. "My thinking was that they were foolish in driving off in that situation," says Thomas. "They didn't know what they were driving into."

But his own situation was scarcely better. With the power out and the fire on their doorstep, the Thomases were cut off and entirely on their own. What they would not find out until much later was that the fire that was racing toward them had already become the deadliest single blaze in Australian history.

The wind shift two hours before had turned the blaze's wall of flame and sent it racing to the northeast, toward the community of Kinglake. Strong wind, steep terrain, and tinder-dry, oily eucalyptus combined to form the deadliest kind of wildfire, an incendiary chain reaction called a blowup. As heat bakes a tree past its flash point, its volatile gases blend with atmospheric oxygen and ignite almost instantaneously, causing the trees to explode in flames. The intensity of the energy released creates a powerful vortex of air that feeds it with fresh oxygen, sucking in cool air and spewing it upward in a chimney that can pierce the stratosphere. The fire exploded up the ridge toward Kinglake at speeds that topped 80 mph.

Hardest hit was a tidy neighborhood of homes along Pine Ridge Road, where a triangle of land was flanked on two sides by steep hillside. Topography that once provided fine views over the southern plain now exposed the community to being overrun by fire from two directions at once. The entire community was caught unawares. There was no time to contemplate the options.

Rob Richings, a service technician, decided to make a run for it once the windows of his house started to explode from the heat. "It's against the rules, but this wasn't a normal bush fire," he later told a reporter. As it was, he managed to drive through the flames and reach safety. Many others did not. Disoriented in the smoke, cars crashed into one another on the jammed road. Flames melted tires and exploded fuel tanks. In one car, six people died together when their vehicle was consumed by flames.

Staying put was just as much a gamble. Another neighbor, Tina Wilson, decided to stay, taking her three children over to the nearby home of Paul and Karen Roland, who were holed up there with their two daughters. "The house has got sprinklers on the roof and we'll be fine," Wilson told her partner over the telephone. "I'll call you soon." Soon after, Karen Roland phoned with her sister. "It's too late!" she yelled over the roar of the fire. "We're trapped!" All nine perished within the burning walls.

Within 30 minutes, the conflagration had passed over the town and moved on. By the time the fire was burning its way through to the Thomas' tree line, 70 people were already dead.

Two minutes after the Thomas' neighbors left their driveway in their attempt to flee Glenburn, the couple called to report that the pasture along the side of the road was on fire. Ian Thomas walked outside. The sky above the tree line was glowing orange. Here and there, falling embers were igniting spot fires. There was little time now. Already he could hear the roar of the approaching flames. Across the street, a line of trees erupted in fire. The fire leapt over the road, each tree igniting the next.

Thomas had counted on his sprinkler system to protect his house and yard from the fire, but the pump was electric, and the power lines were down. For just such a contingency, he had a gasoline-powered generator at the ready. He started it up. Within minutes, the generator's engine coughed and died, and Thomas tried in vain to restart it. His first line of defense was gone. If he and his wife were going to fight the fire, they'd have to do it by hand, with buckets.

The smoke grew so thick that it was impossible to see more than a few feet. Thomas worried that he and his wife would lose contact amid the inky darkness and the deafening roar of the flames. "It was like a steam train coming at you," he says. Soon the fire had surrounded the house, the flames creeping toward them over the grass like a rising tide. From a nearby house came the artillery-like *whuump* of a propane tank exploding.

"I didn't know how things were going to pan out," Thomas says. "It was obviously dangerous. It was very clear that if the house started to go up we would be in real trouble."

Thomas and his wife had committed themselves to their decision. Whether it was the right one, they had no way of knowing. All they could do now was to handle themselves as best they could.

ASSESS

The first step to dealing with a crisis is acceptance. It may sound obvious, but studies of disasters have found that many people remain in denial in the face of evident danger. Nightclub patrons continue to dance and order new rounds of drinks as smoke fills their burning hall; passengers on a sinking ferry sit and smoke cigarettes as the vessel lists ever more ominously to one side. This denial is driven by a mental phenomenon called "normalcy bias." Psychologists say that people who have never experienced a fatal catastrophe have difficulty recognizing that one could be unfolding.

Clearly, this was not a problem for the Thomases. As a professional fire researcher, Ian was all too aware of the risks of conflagration, and he knew his area was highly flammable. He had spent the day checking his home's fire defenses, getting his property physically ready, and preparing himself mentally for the potential danger. Even though the fire's arrival took him by surprise, he wasted not a second accepting the situation for what it was. Some of the residents of Pine Ridge Road in Kinglake had less of a sense of urgency, and so had been less prepared to save themselves.

Assuming that one has avoided denial, the most terrifying part of a crisis is likely to occur at the very beginning, while the full scope of the danger remains unclear. As we've seen, anticipatory fear is often worse than the experience itself. Performers who throw up before every performance never throw up on the stage itself. Personally, I found the scariest part of jumping out of an airplane the instant before I left the door. That's a common experience among first-time skydivers, according to psychologist Seymour Epstein at the University of Massachusetts at Amherst.

Epstein conducted a study in which novice jumpers were fitted with heart-rate monitors that measured their pulses as their plane climbed upward toward its release point. Epstein found that the novices' heart rates got faster and faster until just before they jumped, but once they were out of the plane, their heart rates declined precipitously. The most stressful part of the whole experience, then, was the immediate anticipation. Compared to that, free-falling was a relief: The jumpers finally understood what they were in for.

As a general principle, uncertainty is inherently stressful. Laboratory animals are less stressed by an electric shock when they're warned that it's coming by a buzzer. For humans, uncertainty in the face of danger magnifies stress by forcing a person to think about a wide range of possible outcomes and weigh the possible strategies for dealing with those outcomes. It also allows worst-case scenario thinking that can detract from useful problem-solving. A key early step to combating fear is to find out as much information as possible about the threat at hand.

When we're facing a life-threatening situation for the first time, one of the biggest uncertainties we face is what will happen inside our own minds. The familiar pattern of automatic, habitual responses that we come to think of as essential to our nature gets thrown out the window. Having been in mortal danger before can help a great deal by providing a sense of what one's mind will be like and whether one will hold up under pressure.

When Dave Boon's car was struck by a Class IV avalanche on a highway near Denver, he benefited from having been in another, very different life-threatening situation two years before. He'd been white-water rafting when his boat was swept by the force of a rapid below an overhanging rock.

"It ripped off the oarlocks on both sides of my raft, flattened me out, took the skin off the top of my left knee, and scraped my helmet up pretty bad," he says. "I tried to do a maneuver to spin out, and it didn't work. I was being raked under this wall that was eventually going to put me under water. And I said to myself, 'This isn't good.' And then as I got flatter and flatter, and it was scraping my knee off and my helmet, I said to myself, 'This really isn't good.'"

Boon didn't panic, and the force of the water eventually pulled him free. Two years later, as he found himself tumbling end-over-end inside the avalanche, he knew he wouldn't panic then, either. He knew what his mental process would be like. And that was a powerfully reassuring piece of information.

ACT

We can't do anything if we don't know what's going on. And if we can't do anything, we're helpless. That's an inherently stressful condition. Numerous experiments have shown that being out of control of a negative situation leads to the release of the stress hormone cortisol. Laboratory animals that are unable to prevent themselves from receiving an electrical shock, for instance, become highly stressed. The part of the brain that an animal uses to recognize whether or not it has any control is the ventromedial prefrontal cortex (vmPFC). If it decides that the stressor is uncontrollable, it activates a suite of responses very similar to that of major clinical depression. Depressed individuals seldom have the wherewithal to make proactive decisions and carry them out.

A sense of control, in contrast, will diminish the subjective sensation of stress. In laboratory experiments, subjects are less bothered by electric shocks when they believe that they have some control over them. In general, the more control a person has over a threatening situation, the less anxiety it provokes.

If feeling that you *could* take action is a powerful stress reliever, then actually doing so is even better. Engaged in useful activity, it's easier to stop thinking about your internal experience of fear and instead focus usefully on external things, such as improving your situation. Military psychologists in World War I found that soldiers who were forced to passively endure bombardment with no outlet for useful activity, such as attacking the enemy, were likely to crack under the pressure and become psychological casualties.

Some people are generally more prone to take an active approach in a crisis. Optimists, who tend to foresee positive outcomes, and extraverts, who are socially outgoing, are consistently more proactive. So are people who see themselves as capable of shaping the outcome of whatever situation they find themselves in, a quality that psychologists call an "internal locus of control." A related concept is self-efficacy, a person's belief that she is capable of accomplishing a given task. People with these character traits tend to perceive and take advantage of opportunities to change the situation they find themselves in. Consequently, they cope with stress better. Instead of looking at a glass and thinking that it's half full, they ask: "Where's the faucet?"

Matthew Lieberman's team at the University of California, Los Angeles, conducted a study in 2008 which found that people with these kinds of proactive character traits—which he calls "psychosocial resources"—had lower cortisol levels when asked to give a presentation in front of a group of people. When he put them in a brain scanner and showed them images of frightening faces, he found that they also showed great activation of the right ventral lateral prefrontal cortex (VLPFC), the part of the brain that's central to self-control. This in turn correlated with lower activation of the amygdala.

These are the sorts of people you want with you when the going gets hairy. In 1967, mountain climber Art Davidson and two buddies were trapped in an ice cave near the summit of Denali, Alaska by a raging winter storm. Days went by as they slowly succumbed to hypothermia and starvation, nearly immobile in their tiny hole. They kept themselves going by making careful plans about the only thing they had any control over, their meager supply of rations. When the food ran out, they fell into despair—then managed to find another problem to grapple with: how to locate a cache of fuel that one of them remembered was hidden nearby. By stringing a series of meager hopes together, they managed to survive six days, at which point the weather broke and they escaped down the mountain.

BOND

Almost everyone is more prone to being overwhelmed by fear when they're alone. We're social animals. The community of friends, relatives, and acquaintances also provides an irreplaceable support system that can help mitigate the effects of fear—or indeed almost any kind of stress.

People who find themselves in danger instinctively seek out social support, a phenomenon known as "milling." People cluster together to chat, share information, and basically work to form a consensus about what is going on and what they should do about it. Unfortunately, by then it may well be too late.

On a spring evening in 1977, 1,360 patrons attended a comedy performance at the Beverly Hills Supper Club near Cincinnati, Ohio. The night's main event was about to start when a busboy burst into the room and hurried onto the stage. Taking the mike from one of the comedians, he informed the crowd that a fire had broken out and that they must evacuate immediately. Some of the guests understood at once and began to move toward the exits. Others assumed that the busboy must be part of the comedy act—after all, how often does a building burn down? Others milled. They chatted, speculated, gossiped, even ordered drinks. Smoke began to fill the room. Then, two minutes after the busboy's warning, the room exploded in a fireball. The remaining patrons panicked. Many escaped, but others did not. Ultimately, firefighters found 134 bodies in the room, several of them still seated upright at their dinner tables where they had been enjoying the show.

By distracting patrons from the danger at hand, milling proved deadly during the Beverly Hills fire. But the social bond can also be a lifesaver, particularly in combat. Mutual bonds of trust and friendship hold military units together in situations that would reduce any one man to terror. The rigor of combat itself can turn a group of perfect strangers into, as Shakespeare put it, a "band of brothers." William Manchester wrote in his memoir of fighting in the Pacific during World War II: "Those men on the line were my family, my home. They were closer to me than I can say, closer than any friend had been or would ever be."

Given the importance of social bonds in motivating men to fight, it's interesting that the U.S. military has long forbidden gays to serve openly on the grounds that they would threaten their unit's bond of trust. Plutarch, the Roman historian of the second century AD, held the opposite view, noting that "a band cemented by friendship grounded upon love is never to be broken, and invincible." As evidence for this assertion he cited the example of the Sacred Band of Thebes, a famous unit made up entirely of 150 homosexual couples. As a measure of its cohesiveness in the face of danger, the entire group fought to the death rather than surrender to the invading armies of Philip of Macedon, father of Alexander the Great, in 338 BC. According to Plutarch, when Philip surveyed the battlefield in the aftermath of his victory, he was so moved by the gruesome evidence of their bravery that he burst into tears and declared, "Perish those who suspect those men of doing or enduring anything base." Above their grave he erected a monument that can be seen today.

The homosexuality of the Sacred Band seemed not to have demoralized the rest of the army. On the contrary, as an elite unit, it was often split up into smaller detachments and stationed throughout the front line, to bolster the confidence of the rest. Military commanders have long understood that visible displays of courage can bolster the steadfastness of others. In a survey of Spanish Civil War veterans conducted soon after the end of that conflict, a large majority said that they had "fought better after observing other men behaving calmly in a dangerous situation." Courage can be especially inspiring when demonstrated by a person in a position of leadership.

But if courage is contagious, so too is fear. Lacking leadership or training in high-stress situations, a civilian crowd can easily be stampeded into panic. In nightclub fires like the one that struck the Beverly Hills Supper Club, an initial period of passivity and milling often switches over catastrophically to a contagion of fear, with patrons clogging the exits with their bodies so that no one can escape. That's precisely what happened at the Station nightclub on February 20, 2003. Nearly five hundred people were crammed into the Rhode Island rock 'n' roll venue to hear the band Great White. Soon after the start of the band's first song, pyrotechnics lit by the group's stage manager set fire to acoustic foam that lined the wall behind the stage. Within seconds flames stretched from the floor to the ceiling. The band stopped playing. The lead singer said, "Wow, this ain't good." For a moment the patrons milled about, uncertain what was happening. Then the fire alarm went off, and the crowd surged toward the front door. Gripped by terror, most of them committed a common error: Never having taken the deliberate precaution of noting where the fire exits were, they moved automatically, en masse, toward the door that they had come in through on the way in. In less than a minute, a pile of bodies was jammed in the doorway, many of them half-in, half-out; in sight of safety, yet squeezed and suffocating as thick black smoke rolled over them. Ultimately, 31 bodies were recovered from the front door and the area just inside. None were found by the building's other three exits.

How does fear spread from one person to another? As social animals, we're designed to catch one another's emotions. Empathy is fundamental to our nature. Neuroscientists have identified a specific anatomical feature of the brain, so-called mirror neurons, that fire whether we see someone else perform an action or we perform it ourselves. You could say that we live what others are living via our imagination. When we see someone become embarrassed, we feel embarrassed. When we see someone injured, we feel queasy. The face of a terrified person is itself terrifying.

There may be an even more direct route between one person's fear and another's. When I jumped out of an airplane as part of Lilliane Mujica-Parodi's experiment at Stony Brook University, I wore gauze pads that collected my sweat, as did my fellow skydivers. Later, Mujica-Parodi's team asked other volunteers to breathe extracts of our sweat while they were lying in a brain scanner. Sure enough, their brains showed heightened activation in the amygdala, even though they were conscious of no particular aroma. Human beings, she concludes, seem to release pheromones during stress that send an alarm signal to others in the vicinity.

Take-home lesson: When faced with fear, be careful whom you mill with. Panic and courage are both contagious.

USE EMOTION

When the body is in a state of high arousal, it becomes impossible for the C-system to override the X-system. But one X-system process can override another. The emotions of anger and hatred, in particular, have the power to overwhelm feelings of trepidation. "In a fight," the Welsh proverb says, "anger is as good as courage."

Military leaders have long used the power of anger to rile up their troops and squelch their fear. In their harangues, they might remind an army about the outrages that have been conducted by the enemy, the enemy's brutality, and perhaps even the sexual depredations they would like to inflict on the soldiers' womenfolk. During World War II, when the United States was at war with Japan, the "Nips" were portrayed in American propaganda as brutal, bloodthirsty savages who were both apelike and fiendishly clever. After the war was over, and the Japanese had become steadfast allies of the Unites States, the government quickly shifted the focus to emphasize less threatening aspects of their national character: The Japanese were no longer simian but rather an industrious, polite people with an ardent love of baseball.

It may not be possible to bombard our own minds with propaganda, but we can still put our sense of indignation to work for us in our daily lives. "Anger makes you feel empowered," says Monique Mitchell Turner, a psychologist at the University of Maryland at College Park. When we're bracing ourselves for a difficult confrontation, it can be useful to stoke a grudge rather than forgive and forget.

Stanley Rachman relates the case of an agoraphobic patient, a highly accomplished engineer who began suffering panic attacks that then developed into agoraphobia. He was so paralyzed by fear that he could not leave home without the supportive presence of his wife. The only exception to this extreme dependence occurred during those rare periods when he became very angry with her; then he found that he could leave the house and wander at will. Within a few hours, however, his anger would dissipate, his fear would return, and he would be forced to return home.

Assuming it's appropriate to the occasion, an effective way to get one's blood up is by yelling. The voice has a tremendous psychological effect, both on oneself and on others. Before modern times, it was a given that men marched into battle roaring, chanting, shouting, or singing, both to fortify their own spirits and to terrify the enemy. "The war shout should not be begun till both armies have joined," the

Roman military writer Vegetius advised. "The effect is much greater on the enemy when they find themselves struck at the same instant with the horror of the noise and the points of the weapons."

During the Civil War, Confederate soldiers were famous for their blood-curdling "rebel yell." As the war took place before the advent of audio recording, there's some debate today about what exactly the yell sounded like, but a contemporary described it as a "penetrating, rasping, shrieking, blood-curdling noise that could be heard for miles and whose volume reached the heavens—such an expression as never yet came from the throats of sane men, but from men whom the seething blast of an imaginary hell would not check while the sound lasted."

While the *sound* of yelling has a powerful psychological effect, the mere movement of the breath in and out of the lungs can play an equally important role in keeping emotions in check. Breathing is one of the few processes that is under full control of both the automatic and the reflective parts of the brain, and seems to function as a linkage between them. Every inhalation slightly activates the sympathetic nervous system, while every exhalation activates the parasympathetic. A study conducted by Indian researchers found that students who practiced a yogic slow-breathing technique called *pranayama* for an hour a day for three months exhibited a decline in their resting heart rate—evidence of increased parasympathetic activity, and hence of a lower overall level of stress.

If yoga seems too squishy for you, rest assured that SWAT teams and special forces operators practice breathing skills as well, as part of an overall strategy of physiological control called Tactical Arousal Control Techniques. Four-count tactical breathing is rehearsed until it becomes automatic: breathe in through the nose for a count of four; hold for a count of four; breathe out through the mouth for a count of four; and hold for a count of four. Sandra Glendinning, a police officer in Vancouver, uses "combat breathing" to counter the surge of adrenaline she feels during high-speed pursuits. "I know my adrenaline is going to skyrocket, so before it gets a chance to override my ability to function, I take several slow, deep breaths, completely filling and emptying my lungs," she says. "I 'combat breathe' for as long as it takes for me to gain control over the adrenaline, and until an eerie calm comes over me, enabling me to focus on the task at hand."

Beware: The link between the C-system and the X-system through the breath works in both directions. We've already seen how unconscious rapid breathing can lead to hyperventilation and panic—and how, in some people, deliberate attempts to counter anxiety can trigger relaxation-induced anxiety.

REFRAME

Ultimately, an alligator can't make you scared. A skidding car can't make you scared. The only thing that can make you scared is your mind's interpretation of those things. Fear is a phenomenon that resides entirely within your brain.

Unfortunately, that interpretation lies beyond the control of your conscious thoughts. It's carried out by X-system processes that are effortless and fast, taking place well before you're even aware of the thing in question. In fact, even before the X-system is able to identify what something is, the amygdala has already decided whether or not it's worth being scared of. The X-system's identification happens so quickly, and so far behind the scenes, that it feels like the property of scariness is inherent in the thing itself.

But it's not. And that fact opens the door to the most powerful method of all for controlling fear: Reappraisal. This is the technique of using the C-system to modify how the brain encodes a potential threat. "Change the interpretation, [and] you change your underlying reaction to it," says psychologist Matthew Lieberman. Reappraisal is a natural, instinctive process that we all engage in. "If you see someone in a picture who's all bloody and banged up, and you think to yourself, 'Oh, it's just a Hollywood special effects scene, it's not a real event,' then it's no longer distressing," Lieberman points out.

Some people are better at reappraisal than others. Studies have found that people who are able to think of events as challenging rather than threatening are able to cope better with their emotions, have more positive feelings, and are more confident. Marc Taylor, in his study of

American military personnel undergoing hyper-realistic combat training, found that subjects who relied on positive reappraisal to cope with
their situation had lower levels of stress hormone in their bloodstream
than other personnel.

Contrast that useful kind of positive thinking with the negative appraisal that's common to people in the throes of social anxiety.
Recall that night in 1964 when Sir Laurence Olivier was preparing to
go onstage for his first performance of Ibsen's *The Master Builder.* As he
sensed fear rising within him, his mind jumped forward to all the terrible things that would happen to him if he lost control onstage. The
audience would laugh at him; his friends would abandon him; the critics would savage him. He would never be able to go on stage again: "It
would mean a mystifying and scandalously sudden retirement." In fact,
ever the trooper, he marched on nonetheless, but his vivid imagination
for catastrophe dogged him for five more years.

As we've seen, cognitive behavioral therapy is a powerful tool in
overcoming anxiety disorders, and social anxieties like stage fright in
particular. Patients are taught to recognize when they're thinking unre-
alistically negative thoughts, and to deliberately reassess the situation
in a more positive light. Olivier, for instance, might have reminded
himself that his audience loved him, that he'd been through many such
performances before, and that even if he did "go up" the earth would
still go on revolving. Over time, such thoughts would become second
nature—as thoroughly automatized by the X-system as the negative
thoughts originally were.

Thanks to functional magnetic resonance imaging (fMRI)
brain-scanning technology, it's now possible to watch reappraisal as
it takes place. You'll recall that Columbia University neuroscientist
Kevin Ochsner performed an experiment in which he asked subjects
to look at emotionally charged pictures while lying in a brain scanner. He then asked them to try to control their emotions by thinking
of the image in a way that decreased their response. Sure enough,
these willful attempts at reappraisal activated a region of the prefrontal cortex associated with self-control, while the amygdala was
deactivated.

This mechanism of emotional self-control can exert its effect without a person even being conscious of the fact. Lieberman carried out a follow-up to Ochsner's study in which he asked brain-scan subjects to look at pictures of emotional faces along with two labels that could be used to describe the emotion on the face: angry, scared, happy, surprised. He found that the act of naming the emotion, without any conscious effort on the part of the viewer, was enough to activate the right ventral lateral prefrontal cortex and to reduce activity in the amygdala.

The implication is that merely talking about a negative emotion is enough to undermine its power. This might not be surprising to those who have undergone psychotherapy. Indeed, studies have found that talking or writing about a traumatic experience has a number of health benefits, including improved immune function. One doesn't need to go to a professional therapist. Anyone who's trying to get a grip on their emotions in the heat of a crisis can simply find someone to share their feelings with—or even say them aloud to themselves—in order to regain some control over their X-system.

While actively reflecting on your mental state seems to be positive, mulling over your condition in an aimless way is counterproductive, and in fact harmful to mental health, especially if one is focusing on "what might have been"–type thoughts. Studies have shown that excessive rumination leads to, among other things, a pessimistic outlook and reduced motivation.

Another ineffective approach is to try to suppress your emotions, rather than change them through reappraisal. "Suppression is really not about changing your emotions per se, but about changing the appearance of your emotion," Lieberman says. "It's the grin-and-bear it approach, like when you say, 'I'm not going to let my boss know how mad I am, or let my ex-girlfriend know how sad I am about our breakup. I'm just going to put on my stoic face.' Suppression is good at preventing other people from seeing your emotion, but it's a terrible strategy for actually reducing it. If anything it seems to actually amp it up a bit."

AS THE FIRE RACED toward the Thomases' home, they had no need to suppress their fear. They were too busy taking action. With the pumps gone, they had to fight the fire by hand, plunging buckets of water into emergency cisterns by the house and then hauling them to wherever the danger was greatest.

As the battle began, the Thomases fell into the same camp as novice skydivers in Seymour Epstein's study: They lacked information, having only a vague sense of what defending their home would be like. This early stage, Ian Thomas says, was the most terrifying part of the wildfire. "We simply didn't know what was happening, or what was going to happen," he says.

A hundred feet of lawn ringed the house, forming a firebreak. As the flames crept forward over the grass, Ian and Bronwen shuttled out buckets of water to douse the flames before they could advance too close. Two big pines stood near the house in the front yard, and another in the back yard. If the fire reached any of them, the game would be over. It would be impossible to save the house, and if the house went up, there would be no refuge, no place to survive the heat.

The fire swept through the surrounding trees until it was blazing around them in all four directions. The Thomases worked side by side, except for when a sudden advance somewhere else required one of them to run off to deal with the threat. A series of island-like gardens of native flora stood in the front lawn, and, as the Thomases fought a rearguard action, the gardens ignited one by one into a pillar of flames. "When one went up, it went up with a tremendous rush," says Thomas.

Undermined by the fire, trees began to fall. With a crack, a huge gum tree shuddered and crashed onto their driveway, blocking them in. The fire kept creeping forward, the smoking sea of charcoal inching ever inward behind the front of flame. The Thomases kept patrolling, checking their most vulnerable points, hurriedly lugging buckets of water to counter each new thrust.

Keeping continuously active helped to keep the fear at bay. "We were anxious, but we were just focused on doing what we had to do," Thomas says. As time went on, their growing store of information about the fire also reduced the stressfulness of the crisis. "The longer it went on, in a sense the more comfortable we got with it, because we started to feel that we'd already been to some degree successful, and we stood a chance of continuing to be successful."

Finally, around 2:30 a.m., the situation appeared to stabilize. The fire had crept to within fifteen feet of the front of the house, and within a yard of the back deck, but the flames in the immediate vicinity were now out, and the carpet of burnt-out grass formed a protective barrier. All around them, the carbonized forest glowed with embers and the licking flames of remnant fires. Thomas, nauseous and unsteady from heat exhaustion, could hardly stay on his feet. Together, the weary couple collapsed and slept fitfully for three hours, keeping the blinds open so they could check for flare-ups.

The fight was not over. With the coming of the dawn, the wind began to build, whipping smoldering embers back into flame. Pockets of unburned vegetation erupted like roman candles. Thomas staggered outside to douse the most threatening flare-ups, but he was weak from the night's fight and suffering from heat stroke. He could not take even a sip of water without throwing up. Gradually, the flare-ups became less menacing, and the Thomases began to relax. Except for their house, their property had been incinerated. But they were alive.

The catastrophe of February 7, 2009, dwarfed any of Victoria's previous wildfires. But it was just the beginning. A month later, the Kinglake fire would still be blazing, the fruit of a single spark in a remote hillside pasture that grew into a swath of destruction 50 miles long and 30 miles across. The fire season in Victoria would ultimately claim 210 lives, destroy more than 2,000 homes, and lay waste to a million acres of countryside.

The fires left the people of Victoria wondering whether the "stay or go" policy was to blame for unnecessary deaths. Some argued that the policy should be scrapped in favor of mandatory evacuation.

Thomas disagrees. "I think the policy is the right way to go," he says. "But it's much more complicated than most people think it is. In risk analysis, one of the things you do is try to think of all the possible circumstances that could arise. Being afraid puts you under stress, and that makes it much more difficult to make completely rational decisions. But in the end most people have a very strong survival instinct. They find ways to deal with the situation."

CHAPTER TWELVE

MASTERY

REMEMBER NEIL WILLIAMS, the aerobatic pilot from the beginning of the book who survived a mechanical failure in his airplane thanks to a seemingly impossible feat of quick thinking? It would seem that, given the natural responses of the human central nervous system to imminent danger, such a feat would be impossible. His cognitive faculties should have been utterly squashed by his fear-fueled X-system. So how did he do it? How did he rise to the occasion?

In the intervening chapters, we've delved into the mechanisms of the fear response in some depth, but the mystery doesn't seem any closer to resolution. If anything, Williams' creativity in the face of extreme danger now seems all the more baffling. Perhaps some other factor was at work, some missing psychological attribute. If we look more closely at Williams' story, we might find the clues we need.

IT'S JUST BEFORE NOON on a sunny June day. Isolated clouds waft over the patchwork fields of southwestern England. Williams is practicing a routine that he'll perform at the upcoming World Aerobatic Championships. Today he is in at the airfield with two other members of his aerobatic team, sharing two aerobatic Zlin airplanes between the three of them.

Williams runs through the entire routine twice. Now he's midway through the third iteration. He's just come over the top of a loop and is barreling toward the ground when he pulls back on the stick to pull himself out of the dive and into level flight, like a rollercoaster that's reached the bottom of a steep hill. His high speed, combined with the tightness of his curving path, creates a powerful centrifugal force, or g-force, that squashes him into his seat with five times the force of gravity. He grunts and clenches the muscles in his abdomen to prevent the blood from flowing out of his head and causing him to pass out.

The g-force applies not only to Williams but to the entire aircraft, so that its internal structure is loaded with the equivalent of five times its resting weight—ten thousand pounds, all told. Normally, that wouldn't be a problem; the plane is rated to carry six times its resting weight, or 6 g's. But the plane has seen heavy and frequent use, and the strain has weakened it invisibly. This last maneuver is the straw that breaks the camel's back. Just as the plane returns to level, a structural beam within the left wing collapses.

There is a jolt, a loud *bang*. The aircraft staggers. At once the roar of the wind over the fuselage grows louder, and Williams finds himself thrown sideways against the straps of his harness. Immediately he pulls back the throttle to reduce engine power and minimize aerodynamic strain on the airframe. Then the left wing begins to fold as the plane rolls over and dives toward the earth.

At this point Williams has no idea what is happening. In a flying career that has spanned two decades, he has never heard of such a thing occurring. He is far from the airfield, and quickly losing both altitude and control of the airplane. Instinctively, he pulls back on the control stick, but that only makes the wing fold up further. With three hundred feet remaining between himself and the ground, he loses control of the plane completely. He is in a steep, turning dive. At his current speed, the ground is less than two seconds away. And yet, in the midst of all this, he experiences a rather elaborate thought process.

He remembers (as he will later describe in a written account) that several years before he heard of a Bulgarian pilot who was performing aerobatics in a similar Zlin airplane and had one of its wings break while flying upside down. The plane instantly went out of control and wound up right-side up, at which point the wing snapped back into place. While this scenario was quite different from Williams' own current predicament, he infers a potentially useful principle: In a Zlin with structural wing failure, flipping 180 degrees can drive the wing back into position.

This is an astonishingly evolved rumination for a frightened man with only a few seconds left to live. Yet, in the event, Williams gives the idea a try. He stops fighting the wounded plane's tendency to roll left, and instead pushes it around further to the left until he is completely upside down. Then he moves the stick forward, raising the nose above the horizon. This maneuver subjects him to heavy negative g-forces, so that he hangs upside down in his harness with a multiple of his normal weight. Blood surges down into his face. With another *bang,* the left wing settles back into place.

Williams is not out of the woods—in fact, he's almost literally in the woods, convinced that he's going to plow into a stand of trees. After a few seconds, the plane begins to rise. Soon he is climbing. He begins to think that he might make it, after all. Then the engine sputters to a stop. Another crisis. But Williams has had engines quit on him before. He immediately begins troubleshooting the problem. A check of the fuel-pressure gauge shows that it's reading zero. That narrows the range of possibilities. Next he checks the fuel shut-off valve. It's in the "closed" position. He reasons that he must have accidentally knocked it during the jolt. He turns it back on, and the engine roars back to life. Soon he is climbing again.

So far, so good. He's one thousand feet up, a much safer altitude. Now the crisis becomes: How to get back on the ground? He isn't wearing a parachute. He knows that the plane will quickly become uncontrollable if he tries to fly it right-side up. An upside-down landing would surely kill him instantly. It seems an insoluble conundrum. Worse, he only has eight minutes to ponder it before the plane's inverted gas tank runs dry.

Hanging upside down at one thousand feet, fully aware that he might well die violently in less than ten minutes, Williams is so frightened that he has to brace his knees against the side of the cockpit to keep them from shaking. Yet still he approaches his predicament with the meticulous care of a former test pilot. There are four different ways he can return to earth, he reasons. He can try to land the plane upside down on a pond or lake; try to set down upside-down in a stand of trees, to lessen the impact; crash-land upside down at the airfield; or fly the approach to the airfield upside-down, and then roll it upright at the last minute. The last option he decides, will give him the best chance at survival.

All the same, it's a long shot. The maneuver will require absolute precision and split-second timing. He'll only have one chance to get it right. Rather than despair, Williams sets about narrowing his options further. When he's about to set down, he thinks, he'll have to roll either right or left. Which will be safer? He conducts a short experiment to figure out which direction the plane can roll more safely. He tries rolling to the left, and the wing immediately starts to fold up again. He halts the attempt and gets himself back level again. The last maneuver seems to have weakened the wing even more. It's foolish to carry out any more experiments. He will roll to the right.

Williams comes around the airfield and approaches the landing strip in a shallow descent. Reducing the power to idle, he floats 30 feet above the grass runway until his speed bleeds off to 87 mph, then smartly executes a roll to the right. As the lower wing passes underneath him, its tip slices through the grass of the runway, carving a furrow 36 feet long. As soon as the Zlin is right-side up again, the left wing begins to fold up once more. The result is something in between a crash and a landing, with the plane thumping onto the grass and sliding 70 yards. For a moment after the aircraft comes to a halt, all is still. Then Williams bashes his way through the jammed canopy and runs 20 yards from the wrecked plane. There he collapses with relief.

ANY ONE OF THESE CHALLENGES—the folding wing, the dying engine, the upside-down approach to landing—would have been enough to do in an ordinary pilot. Taken together, their failure to kill Williams challenges our basic understanding of what the human brain is capable of achieving under pressure. But before we give up on trying to understand Williams' feat, let's acknowledge the advantages that he had going for him.

First of all, Williams was extraordinarily well prepared. He was thoroughly habituated, not only to the unusual orientations of aerobatic flight, but to the stress of dangerous in-flight emergencies. He'd had numerous close calls in the course of his several thousand hours' worth of flight time, including one airshow performance that ended with him crashing in front of a crowd of spectators. So being in a life-or-death crisis in an airplane was not a totally new experience for him.

He was also well trained. Indeed, Williams was famously fanatical about honing his skills, and he was constantly educating himself about the aircraft he flew. Discussing the Zlin incident in his book *Aerobatics*, Williams wrote: "The fact that I am here at all is due to my having a detailed knowledge of the structure of that particular aeroplane."

Williams had spent much of his lifetime steeling himself against fear, relentlessly exploring the limits of what an aircraft could do and often finding himself on the wrong end of experiments gone awry. While still a student pilot, he had attempted to pull a loop while out of sight of his instructor. Coming over the top, his inexperience got the better of him, and the plane stalled and began to tumble. Through a combination of luck and nerve, he managed to regain control, and got himself and his airplane safely back onto the ground without his instructor being any the wiser.

Given his familiarity with fear, Williams well understood the mental toll that it could take. "A frightened pilot will never fly as accurately or safely as he would want to," he wrote.

Williams was fundamentally optimistic and proactive—a prototypical internal-locus-of-control person. When his wing first broke he immediately rallied, reminding himself that, as long as he still had some altitude and some aircraft left, there still was plenty for him to do. He started by figuring out how to arrest the collapse of his airplane, then by getting his engine running again, and then climbing to a safe altitude. Once he had time to breathe, he began to assess the situation, using his analytic skills to deduce the condition of his airplane.

Williams was, in short, superb at handling himself in a crisis. But that fact doesn't shed any light on the central mystery: How did he come up with a creative solution under such intense pressure? The brain's X-system, as we know, is fast but reacts reflexively. The C-system, which thinks flexibly and creatively, is very slow. It didn't have the time to consciously reflect on the situation, to sift through its storage banks to come up with an explanation for why his wing was folding up and then extrapolate from that to a plausible solution. Whichever system, then, that was in charge at the moment that Williams' wing fell off—either the C- or X-system—should have proven inadequate to the task at hand.

There's a saying in aviation: A pilot needs to have either airspeed, altitude, or an idea. At three hundred feet off the ground, with his plane moving downward at close to 150 mph, Williams needed an idea and quick.

He got one. It's just not clear from where.

So what part of his brain did Neil Williams use?

IN SEARCH OF AN ANSWER, I travel to Kissimmee, Florida, a town of 60,000 people 12 miles southeast of Disneyworld. Kissimmee blends seamlessly into central Florida's table-flat sprawl of golf courses, bass-fishing lakes, and gated retirement communities. It's also home to something singular: Stallions 51, an outfit that offers flight training and sightseeing rides in World War II–era P-51 Mustang fighters.

Owner Lee Lauderbeck, 58, is an aerobatic pilot with even more experience than Williams had at the time of his crash landing—over seven thousand hours flying Mustangs alone, and some nineteen thousand in total. Even more importantly, as a flight instructor, his primary skill is explaining what's going on and why. If anyone can help me understand what happened to Neil Williams, it's Lauderbeck.

We don flight suits and helmets and climb into the Mustang. Lauderbeck lines us up on the runway and opens the throttle on the huge, 1,700-horsepower, Rolls Royce–built Merlin engines. The

12 cylinders rise to a throaty roar and we start to roll. As we gain speed, the tail lifts, and then we float off the runway. We hold steady, roaring along no more than 25 feet above the ground, as the airspeed indicator passes 150 mph. Then Lauderbeck pulls the stick sharply back and the nose swings up into the blue yonder.

Soon we're barreling along at 200 mph, the sun-dappled Florida flatlands rolling along below us. "Okay," Lauderbeck says, "Your controls." He lifts his hands up toward the canopy so I can see he's not holding the stick.

I make gentle turns to the left and right, just to get the feel of the plane, then Lauderbeck tells me to put the nose down into dive, and we pick up speed until the gauge reads 300 mph. He has me pull back on the stick. I feel heavy in my seat as the nose rears higher, higher, higher. The ground disappears; the canopy in front of me is filled with blue. "That's vertical," he says. "Give me a little bit of right rudder. A little bit more. Just a touch more elevator." Now the horizon is appearing again, the sky and clouds and the brown earth, all the wrong way around. Strangely, I don't *feel* upside down, because the centrifugal force is counteracting gravity and keeping me pinned to my seat.

The horizon slides beneath us, and now we're facing straight down toward the brown-and-green patchwork of central Florida. This is what Williams saw just before his wing broke. It's a lot like being at the top of a roller coaster, except that I don't feel like I'm falling, as the g-forces are holding me into my seat just as if I were sitting flat on the ground.

Now, as our dive bellies, gravity starts to push me down with real weight. We're pulling three g's as we reach the bottom of the loop, a bit more than half of what broke Williams' Zlin. It's a sensation unlike any I've felt since the last time my brother sat on me, probably around the eighth grade. Our wing stays in one piece. Then, as we come through level, the plane shudders, like we've been kicked in the seat. "You've hit your own prop wash," Lauderbeck says: We've come back to the exact spot where we started the loop. "Couldn't have done it better."

Lauderbeck talks me through a half-dozen more maneuvers: a barrel roll; an aileron roll; a half Cuban Eight. I know roughly what I'm doing, but the details are sketchy. It's clear that if something were to go wrong, I'd go tumbling out of control like Neil Williams did when he botched his first loop as a teenager. To become a safe, competent aerobatic pilot, I'd need hundreds of hours of training to burn the correct movements of the controls into my X-system. I would need to overlearn the maneuvers, just as soldiers have been overlearning their military routines for thousands of years.

But there is another level of competence beyond this kind of stock automaticity. Lauderbeck tells me that, after he'd been flying the Mustang for about a thousand hours or so, on top of the ten thousand or so hours he'd put on various jets and helicopters, he began to feel less like he was flying the airplane and more like he was wearing it. "You're not thinking about what you need to do," he says. "It just happens. It's like the plane becomes an extension of your body. You start to feel one with it. It's like you're not getting into it so much as strapping it on."

When he's flying a difficult, zero-margin-of-error maneuver at an airshow—like, say, rolling the plane upside-down when zooming along one hundred feet over the ground—Lauderbeck says he can't really explain what he's doing or how he's doing it. "What do I do with the rudder?" he says. "It just happens. What's the pitch attitude? It just happens. What's the feel of the airplane? It just happens."

At some point, he says, his knowledge of the aircraft, and his familiarity with how it flies, became so integrated into the background workings of his mind that he doesn't need to think about what he's doing at all. "It sounds kind of Star Wars-y to say it," he admits, "But you become one with the aircraft."

THIS, AT LAST, IS THE MISSING part of the equation. Expertise. After thousands of hours behind the controls of an airplane, a pilot like Williams or Lauderbeck has trained far beyond rote overlearning and extended his expertise into a profound kind of understanding. He has reached, in a word, mastery.

Mastery is a heightened level of performance that can be achieved through a combination of talent and, even more importantly, hard work. Years and years of study, and thousands of hours of accumulated practice, can elevate a performer to a level at which his understanding is integrated in a deep, preconscious way.

A substantial literature has been growing up around the subject of what makes experts different from the rest of us. In one of the field's seminal studies, psychologist Adriaan de Groot demonstrated in the mid-1960s that chess players who play at an expert level are able to integrate detailed information about the game at a glance. De Groot briefly showed players chess boards with pieces arranged as if in mid-game, and then asked them to recreate the positions of the pieces on another board. Expert players were able to recall the positions of 91 percent of the pieces, while beginners could only manage 41 percent.

De Groot's conclusion was that the experts had, over years of play, learned to condense disparate pieces of information into "chunks." For example: the queen, threatened by the rook, which is itself threatened by a bishop, but protected by a pawn—all of this information might constitute a single familiar chunk. And so on with the rest of the board. The experts, then, were able to grasp at a glance the meaning of the entire game in front of them in just a handful of chunks. This made the overall situation of the board easy to conceptualize, and hence to remember.

Chunking can allow people to perform seemingly incredible mental feats, by juggling what seems like huge amounts of information. In one imaginative experiment, psychologists paid a university student to practice memorizing increasingly long strings of random numbers. The student practiced an hour a day for two years. By the end of that time, he could memorize strings of up to 80 digits. Few untrained people can manage more than 10. The student's trick was that he had learned to chunk together multiple digits so that he could remember them as a unit. But his overall memory was no better than it had been at the start. Asked to remember a string of words or letters, his was performance was altogether average. The advantages of chunking, it turns out, are limited to the specific set of information on store.

Finding patterns among a swarm of extraneous data is a particular skill of the X-system. Over years of practice, an expert's X-system slowly accumulates, integrates, and stores data about the patterns she encounters in her area of specialty. The X-system can even learn to recognize patterns that the conscious mind is not aware of.

Italian psychologist Cosimo Urgesi conducted an experiment to see how well people could predict whether a basketball shooter would make a basket. His subjects included professional basketball players, sports journalists, coaches, and students who'd never seen or played basketball. He showed each one 12 short clips of a player getting ready to shoot a basket. Each clip ended with the ball in mid-air, and the subjects were asked to predict whether the shot was successful. It turned out that, while everyone had some success at predicting the outcome when they could see the arc of the ball, only the professional players could make accurate calls when the ball had been in flight for just under a second.

Next, Urgesi's team investigated which part of the brain was responsible for the pro player's insight. Using a technology called transcranial magnetic stimulation (TMS), they scanned a region involved in planning motion, the motor cortex, as the players watched the films. It turned out that when the players watched shots that they predicted would miss, an area of the motor cortex corresponding to the little finger became active—specifically, to a muscle in the pinky called the abductor digiti minimi. Though its effect on the trajectory of the ball might seem small, it apparently plays a crucial role in fine-tuning the throw. As they watched the action on film, the professional players were subconsciously reenacting the motion within their own motor cortexes, and using the result to predict whether the shot would succeed. They had no conscious awareness that this was what they were doing; the results surfaced in their consciousness simply as a sense of knowing—a gut instinct of what the outcome would be. Ultimately the results of the prediction would get assimilated into the great pool of data that make up the players' expert knowledge.

All of this subconscious information is encoded in a subregion of the X-system called "procedural memory." We met this piece of machinery before, in the discussion of how the motor system automaticity gets disrupted by choking. It's what underlies our ability to "know how to" rather than to "know what." Unlike explicit memory, which we have conscious access to—for example, where we left our keys—procedural memory is the domain of strictly automatic, subconscious processes, like how to hit a tennis ball or sink a basket. As a person becomes more and more expert at a skill, a process called "knowledge compilation" takes place. The skill turns into an automatic response. It shifts from consciously available explicit memory to the invisible expertise provided by procedural memory.

Not all the information stored inside an expert's brain is inaccessible. Japanese researcher Giyoo Hatano draws a distinction between "routine expertise," which comes from repetitive, rote learning like touch-typing, and "adaptive expertise," which is built up through a more reflective process of problem-solving and discussion. Knowledge acquired this way is accessible to conscious thought; it's the kind of thing experts can talk about. As Lee Lauderbeck said, he can't tell you very much about how exactly he flies a difficult maneuver, but he can tell you everything about the aerodynamics and engineering of the plane he's flying it in. Deep expertise, then, comprises both X-system and C-system knowledge.

There's no quick or easy way to accumulate the vast quantities of information that need to be integrated into an expert's store of knowledge. Psychologist Bill Chase, one of the foremost researchers into the chunking phenomenon, was of the opinion that anyone could learn to be a chess grand master, if they just practiced long enough—ten thousand hours or so, in his estimate.

Having made that enormous time investment up front, an expert attains the gift of lightning-fast performance in the clutch. Psychologist Neil Charness recorded the eye movements of expert and intermediate chess players as they first looked at a chess board with pieces positioned as in the middle of a game. Within the first second, experts' eyes alighted on tactically relevant pieces 80 percent of the time, whereas intermediates managed to do so only 64 percent of the time. Even before they knew what they were looking at, the experts' eyes knew where to go.

Like all well-learned X-system processes, expert performance tends not to decline very much under pressure. Research psychologist Gary Klein conducted a study of expert and intermediate chess players engaged in blitz-style tournaments, in which they were allowed only an average of six seconds per move. Under this time pressure, the intermediates made twice as many bad moves as when they were allowed several minutes per move—whereas the experts actually made slightly *fewer*.

Klein went on to explore how expert decision-makers functioned under even more intense stress. He interviewed firefighting commanders who had spent years battling potentially deadly blazes and found that, when faced with problems like how to get a team into a burning building, they did not consciously deliberate between the pros and cons of various possible options. Instead, they instantly matched the situation to the one most similar in their store of accumulated experience, and chose a solution accordingly. The whole process happens so automatically that one commander that Klein talked to was convinced that he had ESP.

Like master-level chess, then, expertly fighting fires requires a rich store of past experience, organized into a deep intuitive understanding. So does dealing with a plummeting airplane.

ON JANUARY 17, 2009, A US AIRWAYS FLIGHT hit a flock of Canadian geese while climbing out of New York's LaGuardia Airport. The impact of the birds' bodies, traveling at a relative velocity of several hundred miles per hour, instantly shredded the delicate fan blades of the engine. Half a minute later, the pilot in command, Chesley Sullenberger, radioed LaGuardia Tower: "We hit birds. We lost thrust in both engines. We're turning back towards LaGuardia."

In some ways, Sullenberger's position was better than Williams' had been. He still had control of the aircraft, and there were several airports in the vicinity. What's more, the loss of engine power was a crisis that he had trained for specifically in flight simulators. But in other ways, Sullenberger's position was much worse. He was over a densely populated city, with a full load of fuel. And he had 154 other human beings on board with him. A lot of people's lives depended on what he did in the next three minutes.

Listening to recordings of Sullenberger's radio calls during the crisis, what's especially remarkable is how utterly calm he is. He sounds like he's having a routine day. If anything, he might be on the sleepy side of the Yerkes-Dodson curve.

Chalk it up to expertise. At fifty-seven, Sullenberger had accumulated some thirty thousand hours of flying time over a career that spanned more than forty years. He had trained pilots and performed accident investigations. As a sideline, he ran a consulting company that specialized in risk management, and had been appointed as a visiting scholar at University of California, Berkeley's Center for Catastrophic Risk Management. No wasting time in denial for Sully: If ever a person was ready for catastrophe, it was he.

Yet, like Williams, in the moment of crisis Sullenberger found himself gripped by fear. "It was the worst sickening, pit-of-your-stomach, falling-through-the-floor feeling I've ever felt in my life," he told interviewer Katie Couric of *60 Minutes*. "The physiological reaction I had to this was strong, and I had to force myself to use my training and force calm on the situation."

Nosing the plane into a shallow dive to maintain airspeed, Sullenberger had no time for detailed analysis. Like Klein's firefighter commanders, he instantly chose a strategy from his procedural memory: Turn back to the airport. Almost as quickly, he realized that it wouldn't work. He had too little altitude and too far to travel. It just didn't feel right.

Seconds later, the controller asked which runway Sullenberger would like to use at LaGuardia. "We're unable," he replied. "We might end up in the Hudson."

On his second try, Sullenberger selected the course of action that would ultimately turn out to be the best. But it was an awful plan to contemplate: No one had ever before ditched a commercial, wide-body jet without a loss of life. He continued to search for options.

Dead ahead, five miles in from the New Jersey side of the river, lay a small general aviation airport, Teterboro. He asked the air traffic controller about it.

"Do you want to try to go to Teterboro?" the controller asked.

"Yes," Sullenberger said.

The controller called up Teterboro tower and quickly arranged the clearance, then called back to the cockpit: "Turn right 280, you can land at Teterboro."

But in the last 22 seconds Sullenberger's situation had already deteriorated. Teterboro was far, and to make the runway, he'd have to fly in at an angle and then turn sharply. Sullenberger's gut said no. "I can't do it," he radioed. And then: "We're gonna be in the Hudson."

It was his last transmission. While the air traffic controllers kept calling on the radio, trying to offer him options, Sullenberger ignored them. In the final moments before impact, he had entered the cognitive tunnel. His focus was entirely on the problem at hand. "I needed to touch down with the wings exactly level," he explained later. "I needed to touch down with the nose slightly up. I needed to touch down at a descent rate that was survivable. And I needed to touch down just above our minimum flying speed but not below it. And I needed to make all these things happen simultaneously."

A minute later, the plane bellied into the waters of the Hudson River. Its wings were perfectly level. The landing was the first fatality-free ditching of a commercial plane in aviation history. As they coasted to a stop on the icy river, Sullenberger turned to his first officer.

"Well," he said, "that wasn't as bad as I thought."

LOOKING BACK at his own moment of near-catastrophe, Neil Williams explained his moment of insight—turn the plane upside down!—with an elaborate chain of logic that began with a half-remembered story about a Bulgarian pilot and ended up with the decision to invert the plane and pull negative g's. If that explanation seems impossible, it's because it was. Given what we know about the human mind in danger, he couldn't have undertaken such an intricate and explicit thought process.

Instead, the instant that Neil Williams' wing spar snapped, his X-system began integrating all the information about what was going on—what the forces on the stick felt like, what the roar of the slip-stream sounded like, how his stomach seemed to be floating up toward his throat—and matched it against the vast mental library of how the plane he was flying was put together. This database included detailed structural information about how the wing was fastened to the fuse-lage; it also included—perhaps only dimly recalled, but long since integrated into his intuitive model—that story about the Bulgarian pilot. And as his wing folded up, and his adrenaline and cortisol surged toward panic levels, the pattern-matching machinery in his X-system came up with a hit.

Williams did not need to think through to a solution at all. As far as his conscious mind knew, it just came to him, prepackaged in an instant of genius, like a chess master recognizing a checkmate move. Only later, looking back at it with the luxury of time, could Williams infer the logic of what his X-system had carried out. Retrospectively, he believed that he had solved the problem through reflection, but in fact he had acted reflexively.

Fortunately, for the exceptionally well-prepared, reflex can have a kind of genius to it.

CHAPTER THIRTEEN

A NEW CONCEPTION OF COURAGE

WITHOUT FEAR, there would be no need for bravery. It's only when we feel the tide of our emotion rising, when our mind is clouded and it feels like we might lose control, that the necessity of courage comes to the fore. "Courage is fear holding on a minute longer," said George Patton.

One might say that, in the simplest sense, the essence of courage is simply the ability to dominate one's emotions, to stay in charge when terror is trying to wrest away control over your thoughts and actions. To adopt Lieberman's phraseology, it's about the C-system keeping a short leash on the X-system.

But as recent advances have shone a light on how the brain creates and deals with fear, it's becoming clear that the view of courage as a zero-sum game between automaticity and control is not only oversimplified, but potentially dangerous. If the dawning age of neuroscience allows us to gain total mastery of our impulses—if it helps the C-system gain ultimate ascendancy over the X-system—that could turn out to be a very bad thing indeed. To understand why, we need look no further than the crossroads hamlet of Cold Harbor, Virginia.

THE UNION ARMY THAT PANICKED and ran away at the First Battle of Bull Run in 1861 was an army of novices and amateurs, hardly better than the green militiamen that William Morgan deployed at the Battle of Cowpens, or the "raw and undisciplined troops" that Vegetius described as fit for being "dragged to slaughter." They had not habituated to the din and confusion of battle, nor trained automatic X-system processes that could operate in the high arousal state of mortal danger. They were not ready for extreme fear.

In the wake of the disaster, the North realized that the fight was not going to be the quick and easy war it had expected. The South was a determined foe that would not give up until it had been beaten thoroughly into submission. The conflict was going to be long and difficult, and the outcome by no means certain. No more would men be allowed to enlist for 90-day terms. Instead, they would sign up "for the duration." The only way for them to get home would be to fight through to the end.

Methodically, the Union went about creating a massive military machine. Levied en masse, rigorously drilled, and equipped with the latest military technology, the new Union Army swelled from 187,000 to more than one million. More important than the change in size was the change in its nature. Soldiers of the new army were trained to march unflinchingly into mass carnage of a scale that could only be produced by the new industrial age. Troops shuttled about at high speed via the new rail network, shouldering mass-produced rifles and supported by high-velocity cannon. Production, transportation, distribution: The war effort became a vast machine whose product was columns of combat-ready men delivered to the front line.

By 1864, the Union Army was ready to unleash the full power of this force against the enemy. Its new commander, General Ulysses S. Grant, was just the sort of aggressive leader that President Lincoln believed could finally smash the Confederacy. That May, he led the Army of the Potomac on its spring offensive across the Rapidan River and directly into the heart of the Confederacy.

For the next four weeks, he maneuvered his men ever deeper into enemy territory, smashing again and again into the army under the command of General Robert E. Lee. With grim resolve, he battered his way into Virginia in a string of battles that, though technically defeats, left him undaunted. At the Battle of the Wilderness, he lost eighteen thousand men to Lee's eight thousand; at the Battle of Spotsylvania, he lost another eighteen thousand to Lee's twelve thousand. And still he kept on coming.

By early June, Grant had maneuvered his army to within ten miles of the Confederate capital at Richmond, Virginia. At a strategically important crossroads town called Cold Harbor, Lee dug in his army and waited for the assault. On the early morning of June 3, 1864, Lee's 62,000 men were entrenched along a six-mile zigzagging line that ran in a general north-to-south direction. Grant's force, 108,000 men, faced a line of fortifications as elaborate and well-prepared as any yet seen in the war, with multiple lines of trenches and parapets and with concealed gun positions that offered multiple crisscrossing lines of fire.

At 4:30 a.m., Grant launched his attack. Marching through a thick fog toward the Confederate lines, the Union attackers emerged onto the open ground in front of the enemy breastworks and at once began to be shredded by their opponent's massed firepower. Rebel infantrymen, dug in so well that they were nearly invisible, unloaded a hail of lead from their muskets, while their artillery, firing from every angle, delivered canister shot that moved down advancing soldiers like oversized shotgun blasts. Wave after wave marched into the maw of the enemy guns as wounded comrades shot down in previous assaults clutched at their hems, begging them to take cover. But still they marched on, doggedly following orders, knowing full well what lay in store. After the battle, a blood-spattered diary was found on one soldier's body whose final entry read: "June 3, 1864. Cold Harbor. I was killed."

By sunup, some seven thousand Union men lay dead or dying in no man's land. The division commanders decided they could no longer send more soldiers to die. Grant, far from the carnage that he had ordered, realized too late what he had set in motion. Upon hearing the result, he was aghast. "I have always regretted," Grant later wrote, "that the last assault at Cold Harbor was ever made."

That regrettable carnage would not have been possible without the training that had prepared his men for service in a new kind of war. In the face of inevitable slaughter, the Union army was able to march onward, suppressing the X-system that had evolved hundreds of millions of years precisely to prevent this kind of outcome. They had learned to conquer fear. And having done so, they had freed themselves to march into the jaws of their own destruction.

THE HUMAN MIND WAS NOT designed to be ruled by willpower. The C-system evolved within a brain in which the X-system was already well established. The role of the prefrontal cortex added nuance and flexibility to an animal's behavioral suite by guiding, damping down, and sometimes encouraging its more powerful antecessor. It was not designed to function as an all-powerful dictatorship. That's why C-system processes require effort. We don't struggle with our emotions because evolution is inefficient, or because we were designed by a peevish deity. We struggle because self-regulation is *designed* to be difficult.

Over the centuries, human beings have expended a great deal of effort in trying to game their X-systems. We want to control our emotions, and we admire others who succeed. We laud men like Audie Murphy as heroes. We marvel at the tragic fortitude of the three hundred Spartans who volunteered to fight to the death at Thermopylae. We can admire courage even in our enemies. The Roman historian Livy relates the story of Gaius Mucius, a young Roman who infiltrated the camp of the Etruscan army that had laid siege to his city. Failing to assassinate their king, Porsena, he was captured and faced summary execution. Defiantly, he thrust his hand into a fire that was burning on a nearby altar, and showed no emotion as it began to roast. Awed, Porsena let him go.

Such tales of courage, though, carry a dark undercurrent. There's something unsettling about a person who too thoroughly suppresses his own instinct for self-preservation. In 1963, the Vietnamese Buddhist monk Thich Quanh calmly kneeled down in the middle of a busy Saigon intersection, doused himself in gasoline, lit a match, and sat placidly in the lotus position while the flames burned him to death. As his flesh roasted, wrote *New York Times* correspondent David Halberstam, "he never moved a muscle, never uttered a sound, his outward composure in sharp contrast to the wailing people around him." Even if we are sympathetic to his political goal—he was protesting the South Vietnam regime's mistreatment of Buddhists—and awed by his otherworldly composure, it's hard not to feel some horror at his act of self-destruction.

History offers us too many lessons of people who learned to utterly master their fear, and who suffered terribly for it. Before World War II, the Japanese Imperial government increasingly indoctrinated is soldiers with the ideals of the *bushido* code, which romanticized violent death and insisted that "death is as light as a feather." This philosophy greased the country's slide into a disastrous war against China, the United States, and a good portion of Asia; worse, it helped motivate the Japanese to fight in a way that led to maximum ruin and suffering. When the Americans seized the Pacific Island of Saipan in 1944, virtually none of its fifty thousand defenders allowed themselves to be taken alive, preferring to blow themselves up, shoot themselves, or disembowel themselves. Hundreds of women and children jumped off a high cliff to their deaths, one by one. They believed that they were dying honorably. But today, as one watches Japanese come to lay wreaths on Saipan—people who are healthy, prosperous, and free, despite the supposed ignominy of defeat—one can't help but conclude that the willful mastery of fear can be a terrible thing indeed.

WE CAN ERR BY GOING too far the other way, too. We can refuse to engage with fear at all. Our culture, for better or for worse, puts a premium on minimizing risk. Modern technology allows us to avoid many of the dangers that routinely terrorized our ancestors. Perhaps it's no coincidence that so many of us feel a lack of emotional authenticity and connectedness.

Fear is a force that gives life meaning. After Johann Otter was mauled nearly to death by a grizzly bear in Glacier National Park, he underwent a long and painful rehabilitation that included three months with his head wired in a metal halo. Yet, strangely, his experience on that mountain trail is something that now seems indescribably valuable to him. "After something like that, you have a much better realization of the things that are really worth something in life," he says. "Things you can't buy. It's the people around you. It's the people on the trail who found me and kept me alive until I could be rescued. I couldn't have bought their help. It was just good will. That's an incredible realization."

One side effect, Otter says, is that many of life's small daily fears have vanished. He used to loathe public speaking, and was filled with dread every time he was asked to give a presentation at work. Now, he says, "If I have to go in front of a big group, I just think to myself, 'Well, what's worse, facing three hundred people or a grizzly bear?'"

I had a similar experience myself. When I was twelve, my mother was driving me to school on a frosty November morning when she hit a patch of black ice and skidded into a tree at about 35 mph. Neither of us was wearing a seatbelt. I must have blacked out, because I don't remember the impact, or anything, until several minutes later. My mother told me later that after my head slammed into the dashboard I turned to her and a flap of skin fell away from my forehead, and that as I started to talk blood suddenly gushed down my face. So I must have still been conscious. All I can remember is slowly emerging from some kind of an awful dream, a vague and swirling sea of nonsense that was shot through with a profound and disturbing sense that something terrible had happened. Gradually I sensed a pain in my head, and cold. I was surrounded by urgent voices but I couldn't see. I felt confused. I knew I was injured. The idea came to me that I was sitting on a snowbank and I couldn't see.

I felt helpless and very afraid. And I could see my life with a clarity I had never experienced before. It was as though a fog I'd been living in my entire life had lifted, and I saw a truth that I previously missed: that I'd taken everything for granted, and now it might all be gone, forever. I felt an enormous regret. I wish I had lived my life knowing what I knew now.

Gradually I realized that the paramedics were talking to me. They told me I was hurt but that I was okay. A bandage was around my head to stop the bleeding; my eyes weren't damaged. My mother was in pain but her injuries were minor. We rode in an ambulance to the local hospital, where a doctor sewed me up. In time my face healed, but I was never the same. Smashing into a tree was the most transformative event I have ever experienced. At the time, as hurt and terrified as I was, I felt for the first time that I was truly alive, and able to see the truth about existence as I'd never seen it before.

Catastrophe victims frequently report that they lead richer lives in the aftermath. Coming face-to-face with intense danger and overcoming it can help us appreciate our inner strength and renew our gratitude for life. After Hurricane Katrina flooded 80 percent of New Orleans, causing 1,500 deaths and displacing hundreds of thousands, outrage over the scale of the trauma created a national scandal. Yet in the aftermath, eight out of ten survivors said they had a deeper sense of purpose than before the storm.

Experiencing fear can also bring us closer to our friends, families, and partners. In recent decades, an increasing amount of research has been done around a psychological concept called "terror management theory," which posits that people build a sort of psychological buffer around themselves in order to deal with the otherwise unbearable existential anguish that comes from knowing they must die. In a 2002 study, Israeli psychologists found that reminding people of their mortality also made them value their relationships with their loved ones more. Reminders of death, the authors wrote, "heighten the motivation to form and maintain close relationships"; and, conversely, they found that: "The maintenance of close relationships provides a symbolic shield against the terror of death."

The important lesson is not to be afraid of fear, but to work with it. Accept it as part of your life. We should not try to conquer fear, and we should not try to avoid it. The X-system and C-system were built to function as a pair. Each feeds, balances, and counteracts the other. Neither has all of the answers, but if they are working together, we can approximate a mental life that is neither reckless nor circumscribed.

A FINAL NOTE ABOUT Neil Williams. After his miraculous feat of airmanship in landing his crippled Zlin, Queen Elizabeth II awarded him a Commendation for Valuable Service in the Air. The aircraft's manufacturer lent him a replacement, which he used to compete in that year's World Championship. He finished in fifth place. He also continued to compete in the British National Championships, winning it each of the next six times.

His last victory was in 1977. That December, he traveled to Spain to fly a World War II–era Heinkel 111 bomber back from Madrid to the United Kingdom for its new owner. He had flown a similar mission recently, flying another Heinkel back home along the same route for the same purchaser. The legendary German bomber had a reputation for being dangerous to fly, but Williams, at the controls for the first time, could find no grounds for concern: "far from being '*muy peligroso*,'" he wrote, "she was easy and forgiving."

Accompanied by his wife, Lynn, and two engineers, he took off from the Cuatro Vientos airfield on the morning of December 11 and headed north. The sky was partly cloudy, with marginal visibility of about four miles. The plane had no radio navigation equipment that would allow it to fly through clouds, so Williams decided to fly low and follow a road that ran through the Guadarrama mountains north of the city. But conditions were worse in the pass. Fifteen minutes after takeoff, the Heinkel crashed into the side of a cloud-shrouded mountain, killing everyone aboard.

Throughout his life, Williams had lived unfettered by fear. Who can say how much pleasure that freedom gave him, how many things he saw and did that he otherwise never would have experienced? In the end, though, Williams paid a steep price for his courage. Having habituated to danger, he was not particularly motivated to avoid it. He left behind two children, and a legacy to ponder for those who wonder at what the human mind is capable of.

NOTES ON THE SOURCES

Parts of this book appeared in somewhat different form in *Popular Mechanics, Outside's GO,* and *Details* magazines.

INTRODUCTION

Neil Williams

The remarkable story of Williams' crash landing comes largely from his two books, *Airborne* and *Aerobatics*. He also gives a compelling account in an article he wrote for the British magazine *FLIGHT International,* entitled "Structural Failure." Further illumination can be found in the official report of the incident issued by the Accidents Investigation Branch of the UK Department of Trade and Industry.

CHAPTER ONE

Skydiving

As a test subject in Lilly Mujica-Parodi's study, which is still ongoing, I jumped at Skydive Long Island on October 13, 2006. Her work is part of a large body of ongoing research funded by various research agencies within the Department of Defense, which is keen to understand the mechanism of the fear response. Her team has not yet published a paper on the topic that I describe in the chapter, but it has published other papers using the same data, including "Higher body fat percentage is associated with increased cortisol reactivity and impaired cognitive resilience in response to acute emotional stress."

I first described my experience as a subject in this experiment in the *Popular Mechanics* article, "Taking the Scream Test."

The Mystery of Fear

La Rochefoucauld's quote is from his book *Maxims*. The Air Force's attempts during World War II to figure out who would be brave in combat are described in Roy Grinker's book *Men Under Stress*. Rachman describes his study of British bomb disposal experts in his excellent book *Fear and Courage*.

Anatomy of the Fear Response

Joseph LeDoux's speculation on the origin of phobias is contained in his book *The Emotional Brain*, which is an excellent source on the general mechanisms of the brain's fear circuitry, particularly the role of the amygdala. Other enjoyable introductions to the neurological underpinnings of emotion are Antonio Damasio's *Descartes' Error* and Donald Pfaff's *The Neuroscience of Fair Play*. For an overview of how our understanding of the fear response developed historically, see Stanley Finger, *Origins of Neuroscience*.

Also excellent, but more academic in tone, is the overview provided in the first chapter of Phillip M. McCabe, et al., *Stress, Coping, and Cardiovascular Disease*, entitled "Stress Response, Coping, and Cardiovascular Neurobiology." Further insight can be found in Dennis Charney, "Psychobiological Mechanisms of Resilience and Vulnerability" and M. Davis and P. J. Whalen, "The Amygdala."

Finally, an invaluable resource for surveying the state of the art in current research is the NASA report "Stress, Cognition, and Human Performance," by Mark A. Staal.

CHAPTER TWO

Tom Boyle, Jr.

I interviewed Tom Boyle over the telephone on December 15, 2008.

Andy Bolton set the world record for the dead lift at the South East Powerlifting Championships in Eton, England, on April 5, 2009. An article by Michael Davis Smith on the subject can be found at http://www.backporch. fanhouse.com/2009/04/07/andy-bolton-deadlifts-1-008-pounds/.

Yerkes and Dodson's seminal 1908 paper was entitled "The Relation of Strength of Stimulus to the Rapidity of Habit-Formation."

Speed

An entertaining introduction to the startle reflex can be found in the *New Scientist* article "Startle."

For more on how long it takes the brain to respond to a stimulus, see LeDoux's *The Emotional Brain* (page 163). For the slowness of the conscious response, see Daniel Wegner's *The Illusion of Conscious Will* (page 57).

★"Rob Smithee" is not the real name of the Green Beret who told me about his experience in Somalia. As a matter of policy, active duty special forces operators in the U.S. military use pseudonyms when quoted in the press, to protect their identities in case they should ever be captured.

Focus

Samuel Johnson's quote is from Boswell's *Life of Dr. Johnson*. He was referring to the case of the Reverend William Dodd, a popular preacher who was sentenced to death for forgery.

I interviewed David Eagleman over the telephone on December 29, 2008. Since then he has revealed himself as a polymath, publishing a remarkably inventive work of literature entitled *Sum*.

Matthew Lieberman describes the effect of the α2-neuroadrenaline receptor on page 306 of his chapter in the book *Social Neuroscience,* edited by Eddie Harmon-Jones, et al.

Dean Potter's quote is from Jere Longman's *New York Times* article, "900 Feet Up With Nowhere to Go but Down."

The Neil Williams quote is from *Airborne,* pages 173–174.

Memory

Christa McIntyre's quote comes from a press release issued by the University of California, Irvine. Her paper is entitled "Memory-Influencing Intra-Basolateral Amygdala Drug Infusions Modulate Expression of Arc Protein in the Hippocampus."

The value of the "flashbulb memory effect" is described in Joe Z. Tsien's article in the *Scientific American,* "The Memory Code."

The theory that we perceive time based on how many memories we have gets an airing in Caroline Williams' *New Scientist* article "The 25 Hour Day."

Eagleman's research into time dilation was published in the journal *PLoS ONE* under the title "Does Time Really Slow Down during a Frightening Event?" with co-researcher Chess Stetson as the lead author.

Strength

Zatsiorsky's theories about maximal strength are elaborated in his co-authored book, *Science and Practice of Strength Training.*

For an interesting insight into stress-mediated analgesia, see the article by Keith Tully, et al., "Keeping in Check Painful Synapses in Central Amygdala," which discusses the role of noradrenaline in modulating pain. For the role of endocannabinoids, see D. P. Finn, et al., "Evidence for Differential Modulation of Conditioned Aversion and Fear-Conditioned Analgesia by CB1 Receptors." For opioids, see M. S. Fanselow, et al., "Role of Mu and Kappa Opioid Receptors in Conditional Fear-Induced Analgesia." For both cannabinoids and opioids, see R. Butler, et al., "Endocannabinoid-Mediated Enhancement of Fear-Conditioned Analgesia in Rats."

I interviewed Dave Boon on the telephone on August 6, 2007. His story as I relate it combines what he told me with what I read on his Web page, www.daveboon.com.

Going Berserk

A highly entertaining account of *berserkgang* can be found in Howard D. Fabing's paper "On Going Berserk," which contains the quote from Peter Andreas Munch. The story of the Colt .45 draws on Robert A. Fulton's article "The Legend of the Colt .45 Caliber Semi-Automatic Pistol and Moros," which he has posted on his Web site, www.morolandhistory.com, to accompany his book, *MOROLAND 1899–1906: America's First Attempt to Transform an Islamic Society.*

CHAPTER THREE

Task FIGHTER

All information is drawn from the researcher's own paper, Mitchell M. Berkun, et al., "Experimental Studies of Psychological Stress in Man."

Deciphering the Curve

Baddeley's work is described in "Stress, Cognition, and Human Performance," by Mark A. Staal (page 64). Bruce K. Siddle provides an introduction to the practical aspects of the stress response, including the loss of fine motor skills, in his book *Sharpening the Warrior's Edge*. Siddle's book also provided the background for my description of the isosceles shooting stance. My description of the two different kinds of motor control is based on the work of Jean-Alban Rathelot and Peter Strick, as related in their paper "Subdivisions of Primary Motor Cortex Based on Cortico-Motoneuronal Cells." The nineteenth-century dueling guide is *The Art of Dueling,* whose author is given only as "A Traveler."

Running on Auto

Flavius Josephus' appreciation of the Roman Army is contained in Book III, chapter 5 of his work *The Wars of the Jews*. I have used William Whiston's 1864 translation. Philip Lieberman and Stephen Kosslyn describe the neuroanatomy of motor learning on pages 86 and 87 of their book *Human Language and Our Reptilian Brain*. Matthew Lieberman's chapter in *Social Neuroscience* also usefully discusses the acquisition of motor skills. The term "expertise-induced amnesia" was coined by Sian L. Beilock of the University of Chicago, who provides an overview of the topic in her chapter of *Sport and Exercise Psychology: An International Perspective* (pages 154–165).

The Advantage of Doing Things Badly

An intriguing discussion of defecation under stress can be found on page 155 of Neil McNaughton's *Biology and Emotion*. For a classical view, see Victor Davis Hanson's *The Western Way of War* (page 102). The discussion of "sickness behavior" is based on an interview that I conducted with Dr. Wayne Ensign of the U.S. Navy's Space and Naval Systems Warfare Command in October 2006. Easterbrook published his landmark paper, "The Effect of Emotion on Cue Utilization and the Organization of Behavior," in 1959. For an overview of current thinking on the topic, see E. Chajut and D. Algom's paper "Selective Attention Improves Under Stress." The airliner crash in the Dominican Republic was the subject of a safety recommendation by the National Transportation Safety Board (NTSB), "Safety Recommendation A-96-15 through −20." Matthew Lieberman describes the effect of the α1-neuroadrenaline receptor on page 306 of his chapter in the book *Social Neuroscience*. More on "brain fuzz" can be found in H. R. Lieberman et al.'s "The 'Fog of War'" and LeDoux, *The Emotional Brain,* page 285.

A Twelve-Minute Flight

The National Transportation Safety Board issued its full report on the findings of its investigation into the crash, "Aircraft Accident Brief DCA07MA003," on May 1, 2007. Dick Rutan's quote is taken from a review of his instructional video, "Attitude Flying," by Alton K. Marsh in *AOPA Pilot* magazine.

Overwhelmed

Russ Chastain is the moderator of the hunting and shooting forums on *About. com,* where I found his article on buck fever. Joanna Bourke describes methods of snapping civilians out of their panic on pages 250–251 of her book *Fear,* as does Amanda Ripley on pages 132–133 of her book *The Unthinkable.* LeDoux discussed perseveration on page 249 of his book, *The Emotional Brain.* The story of the muskets at Gettysburg is from Dave Grossman's *On Killing* (pages 21–22).

Shutting Down

I first ran across Johann Otter's story in the *Los Angeles Times,* in the form of a two-part article by Thomas Curwen. I then interviewed him over the phone. I also conducted phone interviews with researcher and clinician Charles "Andy" Morgan of Yale University, who is an expert on post-traumatic stress disorder (PTSD) and has conducted research into how elite soldiers respond neurobiologically to intense stress; dissociation is a major predictive factor of future PTSD.

CHAPTER FOUR

Walter Cannon's "fight or flight" idea was explicated in his 1929 book *Bodily Changes in Pain, Hunger, Fear and Rage.* An overview of current thinking on the topic is given in H. Stefan Bracha's article "Does 'Fight or Flight' Need Updating?" I relied heavily on Isaac Marks' superb book *Fears, Phobias, and Rituals* for its clear and entertaining elucidation of the four types of fear response, as outlined in the book's third chapter, "Fear Behaviors: The Four Strategies" (pages 53–82).

Freeze

I interviewed Sue Yellowtail over the telephone on March 19, 2007. Yellowtail is her current married name; at the time of the attack, her name was Groves. For more on the role of the periaqueductal gray in freezing behavior, see M. L. Brandão, et al's "Different Patterns of Freezing Behavior Organized in the Periaqueductal Gray of Rats: Association with Different Types Of Anxiety." I found a helpful discussion of the orienting reflex in Euan M. Macphail's *The Neuroscience of Animal Intelligence* (page 78).

The investigation into Greg Keating's accident was described in the NTSB report NYC06FA205.

Flight

The Pac-Man game study was conducted by Dean Mobbs. Joanna Bourke describes the Bethnal Green Underground Station disaster on page 233 of *Fear.* The Vegetius quote is from his work *De Re Militari*; I used the 1767 translation by John Clarke. An invaluable resource on the Battle of Bull Run is David Detzer's *Donnybrook.*

Fright

David Livingstone recounts his near-fatal encounter with a lion in his book *Missionary Travels and Researches in South Africa*. Marks' description of quiescence is on page 215 of his book. In an interesting follow-up to Cannon's 1942 article, "'Voodoo' Death," University of California at San Diego sociologist David P. Phillips published a paper in 2001 in the journal *BMJ* indicating that people are disproportionately more likely to die of cardiac disease on days that are tradition-ally considered unlucky. I relied on many sources in my retelling of the Virginia Tech shooting story. Among the most useful were Serge F. Kovaleski's *New York Times* article and Clay Violand's own account at WTOP.com.

Fight

The Bible quote is from Joshua 6:21 in the King James Version of the Bible. Patton's quote is from Joanna Bourke's *An Intimate History of Killing*. The Alexis Artwohl quote comes from page 36 of Artwohl and Christenson, *Deadly Force Encounters*. Among the many sources I consulted on the Sean Bell shooting, I found Michael Wilson's *New York Times* article to be the most comprehensive. The Rand Corporation report is entitled *Evaluation of the New York City Police Department Firearm Training and Firearm-Discharge Review Process*. The lead author is Bernard D. Rostker.

CHAPTER FIVE

Cindy Jacobs

I interviewed Cindy Jacobs on January 23, 2009. I found her via the Web site Panic Survivor.com, where I read her account of her struggle with panic disorder.

Panic Disorder

The statistic that panic disorder will affect one in five Americans in the course of their lives comes from page 35 of Antony and Stein, eds., *Oxford Handbook of Anxiety and Related Disorders*, an extremely valuable resource not only for this chapter but also for chapters 6–8. Randi McCabe wrote chapter 23 of the *Oxford Handbook*, "Psychological Treatment of Panic Disorder and Agoraphobia." I interviewed her by telephone on February 5, 2009. Joseph LeDoux discusses the effects of benzodiazepines on the amygdala on page 262 of *The Emotional Brain*. The role of hyperventilation in panic attacks is discussed on pages 36–39 of *Anxiety, Phobias, and Panic* by Reneau Peurifoy and on pages 44–52 of *Panic Disorder* by Richard McNally.

Prevalence

The statistics on the prevalence of panic disorder come from the Web site of the National Institute of Mental Health, www.nimh.nih.gov/health/publications/the-numbers-count-mental-disorders-in-america/index.shtml/. I calculated the figure for percentage of the population using data contained in the United States Census Bureau press release "Census Bureau Estimates Number of Adults, Older People and School-Age Children in States." Sarah Burgard's report is entitled *Perceived Job Insecurity and Worker Health in the United States*.

CHAPTER SIX

Audie Murphy

My primary source for the description of Murphy's combat at Holtzwihr was chapter 19 of his book *To Hell and Back*. For other aspects of his life, I used *Audie Murphy: War Hero and Movie Star*, by Alter and Messersmith, and also relied on the extensive archive of stories on the The Audie Murphy Research Foundation's Web site, www.audiemurphy.com.

The Mammalian Brain

John Allman provides a rollicking account of the evolution of the human brain, including its mammalian aspects, in his book *Evolving Brains*. Jaak Panksepp's book *Affective Neuroscience* covers in great detail the role of oxytocin in social bonding, as well as many other aspects of emotion. The effect of hugging on oxytocin levels was documented by Karen Grewen, et al. in the article "Effects of Partner Support on Resting Oxytocin, Cortisol, Norepinephrine, and Blood Pressure Before and After Warm Partner Contact." The idea that handshaking and other social rituals subconsciously reinforce our status in the social hierarchy is elaborated in Marshall's *Social Phobia*, page 17. Eisenberger's elucidation of a common source for both physical and emotional pain is described in her article "Why Rejection Hurts."

Social Phobia

Jesse Bering's article, "Brave, Stupid and Curious," provides an intriguing look at the phenomenon of "shy bladder syndrome" and the people who study it. A survey of current thinking about social phobia can be found in Stein and Stein's article "Social Anxiety Disorder." The figure that 40 percent of patients don't respond to treatment comes from this article. The origins of erectile dysfunction in social phobia, as well as Masters and Johnson's coining of the term "spectatoring," is from Marshall and Lipsett's *Social Phobia*, page 75.

Peter Kirsch's oxytocin study was entitled "Oxytocin Modulates Neural Circuitry for Social Cognition and Fear in Humans." Maia Szalavitz's excellent *New Scientist* article, " 'Cuddle Chemical' Could Treat Mental Illness," provides an excellent overview of the promise and pitfalls of oxytocin as a remedy for social fear. Stuart Brody reveals another benefit of penetrative intercourse in his article "Blood Pressure Reactivity to Stress Is Better for People Who Recently Had Penile-Vaginal Intercourse than for People Who Had Other or no Sexual Activity." The inclusion of social phobia as a diagnosis within the Diagnostic and Statistical Manual of Mental Disorders (DSM)is described on page 4 of the book *Cognitive-behavioral Group Therapy for Social Phobia* by Richard G. Heimberg.

CHAPTER SEVEN

Dan Jansen

There is a great deal of material available on Jansen, who was a major media phenomenon at the time. (A made-for-TV movie about his life, *A Brother's Promise*, came out in 1996). Particularly useful were Alexander Wolff's 1994 *Sports Illustrated* story; the biography of Jansen on page 3,110 of *Great Athletes*, edited by Johnson; and the biography on ESPN.com by Lisette Hilton.

Choking

John Donovan wrote a very entertaining survey for CNN Sports Illustrated about "chokers" throughout history, "Ankiel, Knoblauch Struggle to Rediscover their Arms." The Steve Blass quote is from Joe Starkey's article "Blass Earns Win over Disease." The Henry Longhurst quote is from an excellent article by Geoff Mangum about the phenomenon of "the yips," entitled "The Neurophysiology of Golf Putting," found at www.puttingzone.com/Dystonia/yipsstudy.html/. The Mackey Sasser story is from pages 136–137 of John R. Marshall's book *Social Phobia*. The characterization of performance anxiety as "nongeneralized social anxiety disorder" can be found on page 116 of David H. Barlow's *Clinical Handbook of Psychological Disorders*. For an overview of the behavioral effects of testosterone, including the "winner's effect," see Panksepp, especially chapter 10. I interviewed John Coates on November 14, 2008. His findings were published in the *PNAS* paper "Endogenous Steroids and Financial Risk Taking on a London Trading Floor." Mackey Sasser's ability to throw normally during on-base play is described in Richard Demak's *Sports Illustrated* article "Mysterious Malady."

The Problem with Thinking

I first encountered Rob Gray's research while reading Mike Stadler's fascinating book *The Psychology of Baseball*. Gray published the results of his research in his paper "Attending to the Execution of a Complex Sensorimotor Skill." Gucciardi and Dimmock published their work under the title "Choking Under Pressure in Sensorimotor Skills."

Trying to Think Less

Wegner introduces the concept of the ironic monitoring process on page 141 of *The Illusion of Conscious Will*. Lieberman and Kesslyn describe how the striatum organizes sequences of action in *Human Language and Our Reptilian Brain*. For a discussion of how motor patterns compete for expression, see Daniel Loach's article, "An Attentional Mechanism for Selecting Appropriate Actions Afforded by Graspable Objects." For more on the globus pallidus and the inhibition of inhibition, see John T. Walkup's *Tourette Syndrome* (page 93). Wegner discusses the "ideomotor effect" on pages 121–125 of *The Illusion of Conscious Will*. Katie Thomas talks about the "double pull" in her *New York Times* article. The section on Tim Woodman is based on an interview with him that I conducted over the telephone on February 7, 2007. For a review of the catastrophe model of choking, see Ivan McNally's article in *Athletic Insight*. For more on holistic word cues as remedies for choking, see Gucciardi and Dimmock's article, cited above.

CHAPTER EIGHT

The Power of Gaze

For the role of the amygdala in facilitating the emotional power of gaze, see Sato et al.'s paper in *Neuroimage*. For Olivier's story I relied primarily on his autobiography and Terry Coleman's biography, *Olivier*.

Freefall

I interviewed Sidney Fortner on February 4, 2009 and Chris Wells on February 18, 2009. The 24 percent figure comes from Martin Fishbein's article in *Medical Problems of Performing Artists*. Sian Beilock's article discussing the role of working memory in choking is entitled "Math Performance in Stressful Situations."

In Bocca al Lupo

Mark Leary discusses famous performers who have suffered from stage fright on page 75 of his book *Social Anxiety*. The statistic that 30 percent of musicians with stage fright eventually stop performing is from Linda Hamilton's *The Person Behind the Mask* (page 56).

Drugs

Hanson explores the role of drinking in Greek warfare in chapter 11 of *The Western Way of War*. Philip Freeman discusses the Celts going naked into battle on page 105 of *The Philosopher and the Druids*. John Koch reviews the importance of alcohol in Celtic warrior culture on page 616 of *Celtic Culture,* noting: "The chieftain's hospitality and ability to provide his warriors with enough alcohol to keep them pleasantly drunk for a year was seen as a testament to his worthiness as a leader." The information on grog in the Royal Navy is from *A Mariner's Miscellany* by Peter H. Spectre. The brain-scan study about alcohol and the amygdala is from Jodi Gilman et al.'s "Why We Like to Drink." A review of barbiturate history, and some of its high-profile victims, can be found in *Molecules that Changed the World,* by K. C. Nicolaou and Tamsyn Montagnon, which also covers the story of beta blockers and their discovery. For a discussion of the hazards of benzodiazepines, see Marshall and Lipsett, *Social Phobia* (pages 189–191). Michael Caine discussed his experience of working with Olivier during his benzodiazepine phase in an interview that was published on the Web site IndieLondon. The *New York Times* article I mention is "Better Playing Through Chemistry," by Blair Tindall; this is the source of the figure that 70 percent of musicians using a beta-blocker obtained the medication from a friend rather than a doctor. Matt Scott wrote the Kim Jong-su story for the *Guardian* (UK). For a contrarian view on efforts to crack down on beta-blocker use, see Carl Elliott's *Atlantic Monthly* piece. I interviewed Linda Hamilton on February 26, 2009.

CHAPTER NINE

The de Montaigne quote can be found on page 365 of the George Bell (1908) edition of his essays.

Antarctica

Information about Russia's Antarctic bases comes from the Web site of the Russian Antarctic Expedition, www.aari.aq/default_en.html. I had heard rumors several years ago about a Soviet physician who had performed an appendectomy on himself in Antarctica, but it wasn't until another physician stationed in Antarctica, South African Ross Hofmeyr, tracked down the story, and posted the details on the Internet, that I was able to find out the details.

X- and C- System

I interviewed Matthew Lieberman on March 12, 2009. Much of the material here is adapted from his chapter, "The X- and C-systems," in *Social Neuroscience,* edited by Eddie Harmon-Jones and Piotr Winkielman. Kevin Ochsner describes his experiment involving the ventral lateral prefrontal cortex (VLPFC) in his own chapter in the same book, "How Thinking Controls Feeling." For further detail on the subject, see Jessica Cohen and Matthew Lieberman's "The Common Neural Basis of Exerting Self-Control in Multiple Domains." Muraven and Baumeister's article is entitled "Self-Regulation and Depletion of Limited Resources." For more on how the unconscious mind can detect patterns without a person being conscious of it, see the Bechara et al. article and Wilson's *Strangers to Ourselves* (page 62).

Failures of Self-Control

Amanda Ripley writes engagingly about the importance of mental preparation before disaster strikes—see pages 126 and 177 of *The Unthinkable*. On the subject of scuba panic, in addition to the articles authored by William Morgan himself, I also found Phil Davis' interview with him, "The Risks of Panic in Scuba Diving," to be very helpful.

Scuba Panic

The Rolf Adams story is largely based on Craig Vetter's 1994 *Outside* magazine article about Bill Stone's Sistema Huautla expedition. Excellent overviews of the subject of scuba panic are provided by John Francis' article in *Scuba Diving* magazine and the Colvards' article in the *Undersea Journal*. Rohrmann's paper is entitled "Changing psychobiological stress reactions by manipulating cognitive processes."

CHAPTER TEN

USS *Trayer*

My experience aboard the simulator was the subject of my *Popular Mechanics* article "The Unluckiest Ship in the Navy."

Habituate

Joseph LeDoux describes how fears can reemerge on page 251 of *The Emotional Brain*. The "mental conditioning chamber" is described by Wendy Rejan in her article, "Shooting Without Bullets." The Carl von Clausewitz quote is from Hancock and Szalma's *Performance Under Stress* (page 273). The comparative sensitization of rural dwellers to bombing during the Blitz is from Rachman, *Fear and Courage* (page 29). For the use of habituation in conquering phobias, see Sam Lubell's *New York Times* article, "On the Therapist's Couch, a Jolt of Virtual Reality." Thomas Joiner's hypothesis is examined in Robert Pool's *New Scientist* article, "Why Do People Die That Way?"

Train

Vegetius Renatus' quotes are from a translation of *De Re Militari*. For a discussion of whether it is best to train with or without stress, see page 59 of Mark Staal's *Stress, Cognition, and Human Performance*. The quote from the skydiving instructor comes from Laura A. Thompson et al., "Context-Dependent Memory under Stressful Conditions." I interviewed Marc Taylor on January 31, 2007. Beilock's study, in which she trained golfers not to choke, was described in her 2001 article "On the Fragility of Skilled Performance: What Governs Choking under Pressure?" For a discussion of the importance of self-efficacy in reducing stress, see page 98 of Helmus and Glenn, *Steeling the Mind*, and page 32 of Jennifer Kavanagh's *Stress and Performance*.

Steel Yourself

Rachman describes his study of British bomb-disposal experts on pages 302–306 of *Fear and Courage*. Mujica-Parodi and Taylor's study was published as "Higher Body Fat Percentage Is Associated with Increased Cortisol Reactivity and Impaired Cognitive Resilience in Response to Acute Emotional Stress." The lead author on Morgan's wetsuit temperature-stress study was K. F. Koltyn and the title of the article is "Influence of Wet Suit Wear on Anxiety Responses to Underwater Exercise."

Prepare to Lose Your Mind

The acknowledged authorities on the subject of heuristics are Nobel Prize winner Daniel Kahneman and his partner Amos Tversky. For the story of the checklist, I relied on Atul Gawande's 2007 *New Yorker* article. A compelling account of the Battle of Cowpens can be found in chapter 91 of Sydney George Fisher's *The Struggle for American Independence*.

CHAPTER ELEVEN

Victoria Burning

A vivid account of the brush fire story can be found in Richard Spencer's article for the *Telegraph* (UK), "Australian Bush Fires' Street Of Death." For a discussion of the mechanics of a raging wildfire, see Domingos Xavier Viegas' article "Anatomy of a Blow-Up." I interviewed Ian Thomas on March 12, 2009.

Assess

Amanda Ripley discusses normalcy bias on page 9 of *The Unthinkable.* The lead author on Seymour Epstein's skydive study was Walter Fenz. For the anxiolytic effects of foreknowledge, see Staal's *Stress, Cognition, and Human Performance*, page 26, and page 20 of Gilbert's *Stumbling on Happiness.* Kavanagh raises the issue of "worst-case scenario" thinking on page 33 of *Stress and Performance.*

Act

For the role of the ventromedial prefrontal cortex (vmPFC) in determining whether a stressor is controllable, see J. Amat, et al., "Medial Prefrontal Cortex Determines How Stressor Controllability Affects Behavior and Dorsal Raphe Nucleus." For the anxiolytic effects of control, see Staal's *Stress, Cognition, and Human Performance*, pages 25–26; as well as page 253 of Muravan and Baumeister's "Self-Regulation and Depletion of Limited Resources: Does Self-Control Resemble a Muscle?" Bourke writes about the concerns of World War I military psychologists on page 205 of *Fear: A Cultural History.* The lead author of Lieberman et al.'s "psychosocial resources" article, "Neural Bases of Moderation of Cortisol Stress Response by Psychosocial Resources," is Shelley E. Taylor. Art Davidson tells the story of his survival atop Denali in his book, *Minus 148 Degrees.*

Bond

A leading expert on disaster psychology, Dennis S. Mileti, discusses the phenomenon of milling in his co-authored article, "The Role of Searching in Shaping Reactions to Earthquake Risk Information." Among the useful sources on the Beverly Hills fire is a special section that the *Cincinnati Enquirer* put out on the twentieth anniversary of the event, "The Beverly Hills Fire: Tragedy Rooted in Code Violations." Shakespeare's "band of brothers" line is from *Henry V,* Act 4 scene 3. Manchester is quoted on page 26 of Hanson's *The Western Way of War.* I used the Project Gutenberg e-book edition of *Plutarch's Lives,* Volume II, available at history-world.org/plut2.pdf, translated by Aubrey Stewart. The monument over the graves of the Sacred Band can be seen on the Jazzhaven Gallery Web site at www.gallery.jazzhaven.com/lion-monument-6365i.htm. The survey of Spanish Civil War veterans is from Rachman's *Fear and Courage,* page 56. The lead author of the report on the investigation into The Station nightclub fire was W. L. Grosshandler. Sandra Blakeslee's article in the *New York Times* offers a compelling description of the state of current knowledge about mirror neurons. Mujica-Parodi's paper on her pheromone research is "Second-Hand Stress."

Use Emotion

A fascinating look at the rapid reversal of America's racial attitude toward the Japanese is provided in Naoko Shibusawa's book, *America's Geisha Ally.* The Monique Mitchell Turner quote is from Dan Jones and Alison Motluk's article for the *New Scientist,* "Eight Ways to Get Exactly What You Want." The story about the agoraphobic patient who overcame his condition when angry at his wife is from Rachman's *Fear and Courage,* page 57. The "Rebel Yell" section is based mostly on Troiani, *Don Troiani's American Battles,* page 218. The pranayama study is G. K. Pal's "Breathing Exercises on Autonomic Functions in Normal Human Volunteers." Sarah Glendinning's description of her use of combat breathing control is from her blog, "Behind the Blue Line."

Reframe

The finding that people who are able to think of events as "challenging" rather than "threatening" are able to cope better with their emotions is from Mark Staal's *Stress, Cognition, and Human Performance,* page 22. For Ochsner's study, see page 120 of Harmon-Jones and Winkielman, eds., *Social Neuroscience,* as well as page 17 of Cohen and Lieberman's "The Common Neural Basis of Exerting Self-Control in Multiple Domains." Lieberman's follow-up study is described in "Putting Feelings Into Words: Affective Labeling Disrupts Amygdala Activity in Response to Affective Stimuli." The benefits of speaking or writing about traumatic experiences can be found in Baumeister's "How Emotion Shapes Behavior." For more on the negative consequences of ruminating, see Sonja Lyubomirsky's co-authored article on the topic, as well as Matthew Lieberman's chapter, "Why Symbolic Processing of Affect" Can Disrupt Negative Affect, in Todorov and Prentice, eds., *Social Neuroscience.* For more on the drawbacks of suppression, see Philippe Goldin, et al., "The Neural Bases of Emotion Regulation: Reappraisal and Suppression of Negative Emotion."

CHAPTER TWELVE

Neil Williams Redux

The sources are the same as in the Introduction. The quote about his "detailed knowledge of the structure" of his airplane is from his *Aerobatics*, page 88.

Stallions 51

My experience flying with Lee Lauderbeck was also documented in an article in the August 2009 issue of *Popular Mechanics*. I flew with Lauderbeck on October 13, 2008.

Expertise

De Groot's chess study is from Eysenck and Keane, *Cognitive Psychology* (page 452). The story of the student who memorized long lists of numbers is from the same book (pages 460–461). I came across the Cosimo Urgesi study on the blog "Neurophilosophy," which is available here: www.scienceblogs.com/neurophilosophy/2008/08/the_ballers_brain.php. The lead author for the study was S. M. Aglioti, and the title of the article is "Action Anticipation and Motor Resonance in Elite Basketball Players." The material about Giyoo Hatano is from Kim Kirsner and Speelman, *Implicit and Explicit Mental Processes* (pages 145–146). The Bill Chase information is from Eysenck and Keane (page 452). The Charness study is from page 453 of the same book. Klein's blitz-chess experiment is described on page 164 of his book *Sources of Power*, while the firefighting study is from page 31.

Sullenberger

I relied on many newspaper accounts of the US Airways ditching. Particularly useful were the FAA memo providing a complete transcript of radio transmissions during the incident (available here: http://www.faa.gov/data_research/accident_incident/1549/), the Associated Press timeline, and the CBS News article describing Katie Couric's interview with Sullenberger.

CHAPTER THIRTEEN

Cold Harbor, Virginia

For my account I relied primarily on Ernest Furgurson's entertaining book, *Not War But Murder*.

The Overdominant C-system

Livy's story is from his *History of Rome, Vol 1*. Halberstam's account of Thich Quanh's self-immolation is from his book *The Making of a Quagmire*. For more on the mass suicide on Saipan, see chapter 14, "Suicide Cliff and Banzai Cliff," in *D-Day in the Pacific* by Harold J. Goldberg.

Fear Gives Life Meaning

The Israeli paper I refer to is Mikulincer et al.'s "The Existential Function of Close Relationships: Introducing Death Into the Science of Love."

Neil Williams Redux Part 2

Williams' description of the Heinkel as "easy and forgiving" is from *Airborne* (page 154). The final chapter in Williams' life was related in a *FLIGHT International* article dated June 24, 1978.

BIBLIOGRAPHY

A Traveler. *The Art of Dueling*. London: Joseph Thomas, 1836.

Accidents Investigation Branch. *Moravan Zlin Z526A G-AWAR Report on the accident at Hullavigton, Wiltshire on 3 June 1970*. Civil Accident Report No. EW/C348/01. London: Her Majesty's Stationery Office, 1971.

Aglioti, S. M., Cesari P., Romani, M. and Urgesi, C. "Action Anticipation and Motor Resonance in Elite Basketball Players." *Nature Neuroscience* 11 (2008): 1109–1116.

Allman, John. *Evolving Brains*. New York: Scientific American Library, 1999.

Alter, Judy, and Patrick Messersmith. *Audie Murphy: War Hero and Movie Star*. Abilene, TX: State House Press, 2007.

Amat, J., Baratta M.V., Paul E., Bland S.T., Watkins L.R. and Maier S.F. "Medial Prefrontal Cortex Determines How Stressor Controllability Affects Behavior and Dorsal Raphe Nucleus." *Nature Neuroscience* 8, no. 3 (March 2005): 365–371.

Antony, Martin M., and Murray B. Stein, eds. *Oxford Handbook of Anxiety and Related Disorders*. New York: Oxford University Press, 2009.

Artwohl, Alexis, and Loren W. Christenson. *Deadly Force Encounters*. Boulder, CO: Paladin Press, 1997.

Associated Press. "Timeline Released of US Airways Flight 1549." *NJ.com*, January 17, 2009. http://www.nj.com/hudson/index.ssf/2009/01/timeline_released_of_us_airway.html/.

The Audie Murphy Research Foundation. "Audie L. Murphy Memorial Web Site." *audiemurphy.com*. http://www.audiemurphy.com/.

Barlow, David H. *Clinical Handbook of Psychological Disorders*. New York: Guilford Press, 2001.

Baumeister, Roy F., Vohs, K. D., DeWall, N., Zhang L. "How Emotion Shapes Behavior." *Personality and Social Psychology Review* 11, no. 2 (2007): 167–203.

Bechara, Antoine, Damasio, H., Tranel, D., and Damasio, A. R. "Deciding Advantageously Before Knowing the Advantageous Strategy." *Science* 275 (February 28, 1997): 1293–1295.

Beilock, S. L. "Math Performance in Stressful Situations." *Current Directions in Psychological Science* 17, no. 5 (2008): 339–343.

Beilock, S. L., and T. H. Carr. "On the Fragility of Skilled Performance: What Governs Choking under Pressure?" *Journal of Experimental Psychology: General* 139, no. 4 (2001): 701–725.

Bering, Jesse. "Brave, Stupid and Curious: Dangerous Psychology Experiments from the Past." ScientificAmerican.com, January 13, 2009. http://www.scientificamerican.com/article.cfm?id=brave-stupid-and-curious/.

Berkun, Mitchell M., Bialek, H. M., Kern, R. P., and Yagi, K. "Experimental Studies of Psychological Stress in Man." *Psychological Monographs: General and Applied* 76, no. 15 (1962): 1–39.

Blakeslee, Sandra. "Cells That Read Minds." *New York Times,* January 10, 2006.

Boon, David, telephone interview with author, August 6, 2007.

———. "Swept Away in an Avalanche!" Daveboon.com. http://www.dave-boon.com/newsletters/Dave_Boon-Swept_Away.pdf.

Bourke, Joanna. *Fear: A Cultural History.* Emeryville, CA: Shoemaker & Hoard, 2006.

———. *An Intimate History of Killing: Face-To-Face Killing in Twentieth-Century Warfare.* New York: Basic Books, 2000.

Boyle, Tom, telephone interview with the author, December 15, 2008.

Bracha, H. S., Ralston, T. C., Matsukawa, J. M., Williams, A. E., and Bracha, A. S. "Does 'Fight or Flight' Need Updating?" *Psychosomatics* 45 (October 2004): 448–449.

Brandão, M. L., Zanoveli J. M., Ruiz-Martinez R. C., Oliveira L. C., and Landeira-Fernandez J. "Different Patterns of Freezing Behavior Organized in the Periaqueductal Gray of Rats: Association with Different Types Of Anxiety." *Behavioral Brain Research* 188 (March 17, 2008):1–13.

Brody, Stuart. "Blood Pressure Reactivity to Stress Is Better for People Who Recently Had Penile–Vaginal Intercourse than for People Who Had Other or no Sexual Activity." *Biological Psychology* 71, no. 2 (February 2006): 214–222.

Burgard, Sarah, Jennie Brand, and James S. House. *Perceived Job Insecurity and Worker Health in the United States.* Professional Standards Commission Research Report, no. 08–650 (July 2008).

Butler, R. et al. "Endocannabinoid-Mediated Enhancement of Fear-Conditioned Analgesia in Rats: Opioid Receptor Dependency and Molecular Correlates." *Pain* 140, no. 3: 491–500.

Cannon, Walter B. *Bodily Changes in Pain, Hunger, Fear and Rage: An Account of Recent Research into the Function of Emotional Excitement*, 2nd ed. New York: Appleton-Century-Crofts, 1929.

———. " 'Voodoo' Death." *American Anthropologist* 44, no. 2 (1942).

Carter, Sue C. "Neuroendocrine Perspectives on Social Attachment and Love." *Psychoneuroendocrinology* 23, no. 8 (November 1998): 779–818.

CBS News. "Flight 1549: A Routine Takeoff Turns Ugly." *CBSNews.com*, February 8, 2009. http://www.cbsnews.com/stories/2009/02/08/60minutes/main4783580.shtml.

Chajut, E. and D. Algom. "Selective Attention Improves Under Stress: Implications for Theories of Social Cognition." *Journal of Personality and Social Psychology* 85, no. 2 (2003): 231–248.

Chastain, Russ. "The Hunter's Autopilot: When our Subconscious Saves Us from Ourselves." *About.com.* http://hunting.about.com/od/deerbiggame/a/autopilot.htm/.

Charney, Dennis S. "Psychobiological Mechanisms of Resilience and Vulnerability: Implications for Successful Adaptation to Extreme Stress." *American Journal of Psychiatry* 161, no. 2 (February 2004): 195–216.

Cincinnati Enquirer. 1997. "The Beverly Hills Fire: Tragedy Rooted in Code Violations." May 28. Available at http://www.enquirer.com/beverlyhills/index2.html.

Coates, John, telephone interview with the author, November 14, 2008.

Coates, J. M., and J. Herbert. "Endogenous Steroids and Financial Risk Taking on a London Trading Floor." *Proceedings of the National Academy of Science* (PNAS) 104, no. 16 (April 22, 2008): 6167–6172.

Cohen, Jessica R., and Matthew D. Lieberman. "The Common Neural Basis of Exerting Self-Control in Multiple Domains." Forthcoming in Y. Trope, R. Hassin, & K. N. Ochsner (eds.) Self-*control*.

Coleman, Terry. *Olivier*. New York: Macmillan, 2006.

Colvard, David F., and Lynn Y. Colvard. "A Study of Panic in Recreational Scuba Divers." *Undersea Journal*, First Quarter 2003. Available at http://www.divepsych.com/UJ1Q03p040_044_qxd.pdf.

Curwen, Thomas. "Attacked by a Grizzly." *Los Angeles Times*, April 29, 2007.

Damasio, Antonio. *Descartes' Error*. London: Penguin Books, 1994.

Davidson, Art. *Minus 148 Degrees*. Seattle, WA: Mountaineers Books, 1999.

Davis, M., and P. J. Whalen. "The Amygdala: Vigilance and Emotion." *Molecular Psychiatry* 6 (2001): 13–24.

Davis, Phil. "The Risks of Panic in Scuba Diving." *seagrant.wisc.edu*, October 23, 2002. http://www.seagrant.wisc.edu/Communications/diving/panicq&a.htm.

Demak, Richard. "Mysterious Malady." *Sports Illustrated*, April 8, 1991.

de Montaigne, Michel, ed. by W. Carew Hazlitt. "The Essays of Michel de Montaigne." London: George Bell, 1908.

Detzer, David. *Donnybrook: The Battle of Bull Run, 1861.* New York: Houghton Mifflin Harcourt, 2004.

Donovan, John. "Ankiel, Knoblauch Struggle to Rediscover their Arms." *CNNSI.com,* March 23, 2001. http://sportsillustrated.cnn.com/baseball/mlb/2001/spring_training/news/2001/03/23/ankiel_knob/.

Eagleman, David, telephone interview with the author, December 29, 2008.

Easterbrook, J. A. "The Effect of Emotion on Cue Utilization and the Organization of Behavior." *Psychological Review* 66 (1959): 183–201.

Eisenberger, Naomi I., and Matthew D. Lieberman. "Why Rejection Hurts: A Common Neural Alarm System for Physical and Social Pain." *Trends in Cognitive Sciences* 8 no. 7 (July 2004): 294–300.

Elliot, Carl. "In Defense of the Beta Blocker." *Atlantic Monthly,* August 20, 2008.

Ensign, Wayne, telephone interview with the author, October 24, 2006.

Eysenck, Michael W., and Mark T. Keane. *Cognitive Psychology.* London: Taylor & Francis, 2005.

Fabing, Howard D. "On Going Berserk: A Neurochemical Inquiry." *Scientific Monthly* 83, no. 5 (Nov. 1956): 232–237.

Fanselow, M. S., Calcagnetti. D. J., and Helmstetter, F. J. "Role of Mu and Kappa Opioid Receptors in Conditional Fear-Induced Analgesia: The Antagonistic Actions of Nor-binaltorphimine and the Cyclic Somatostatin Octapeptide, Cys2Tyr3Orn5Pen7-amide." *Journal of Pharmacology and Experimental Therapeutics* 250, no. 3 (September 1, 1989): 825–830.

Federal Aviation Administration. *Memorandum Date: February 2, 2009 To: Aircraft Accident File N90-TRACON-0122. From: New York Terminal Radar Approach Control Facility. Subject: INFORMATION: Full Transcript Aircraft Accident, AWE1549.* New York New York, January 15, 2009.

———. "USAirways 1549 (AWE1549), January 15, 2009." *FAA.gov,* March 2, 2009. http://www.faa.gov/data_research/accident_incident/1549/.

Fenz, Walter D., and Seymour Epstein. "Gradients of Physiological Arousal in Parachutists as a Function of an Approaching Jump." *Psychosomatic Medicine* 29 (1967): 33–51.

Finger, Stanley. *Origins of Neuroscience.* New York: Oxford University Press, 1994.

Finn, D. P., Beckett, S. R., Richardson, D., Kendall, D. A., Marsden, C. A., and Chapman, V. "Evidence for Differential Modulation of Conditioned Aversion and Fear-Conditioned Analgesia by CB1 Receptors." *European Journal of Neuroscience* 20, no. 3 (August 2004): 848–852.

Fishbein, Martin, Middlestadt, S. E., Ottati, V., Straus, S., and Ellis, A. "Medical Problems Among ICSOM Musicians: Overview of a National Survey." *Medical Problems of Performing Artists* 3, no. 1 (March 1988): 1.

Fisher, Sydney George. *The Struggle for American Independence.* Philadelphia: J. B. Lippincott, 1908.

FLIGHT International. 1978. "AIB Reports the Neil Williams Accident." June 24.

Fortner, Sidney, telephone interview with the author, February 4, 2009.

Francis, John. "Never Panic Again." *Scuba Diving,* August 2005.

Freeman, Philip. *The Philosopher and the Druids.* New York: Simon and Schuster, 2006.

Fulton, Robert A. *MOROLAND 1899–1906: America's First Attempt to Transform an Islamic Society.* Tumalo Creek Press, 2007.

———. "The Legend of the Colt .45 Caliber Semi-Automatic Pistol and Moros." *Morolandhistory.com.* http://www.morolandhistory.com/.

Furgurson, Ernest B. *Not War But Murder: Cold Harbor 1864.* New York: Alfred A. Knopf, 2000.

Gawande, Atul. "The Checklist." *New Yorker,* December 10, 2007.

Gilbert, Daniel. *Stumbling on Happiness.* New York: Random House, 2007.

Gilman, Jodi M., et al. "Why We Like to Drink: A Functional Magnetic Resonance Imaging Study of the Rewarding and Anxiolytic Effects of Alcohol." *Journal of Neuroscience* 28 (April 30, 2008): 4583–4591.

Glendinning, Sarah. "Combat Breathing." *Behindtheblueline.ca,* January 21, 2009. http://www.behindtheblueline.ca/blog/blueline/tag/tactical-arousal-control-techniques/.

Goldberg, Harold J. *D-Day in the Pacific: The Battle of Saipan.* Bloomington: Indiana University Press, 2007.

Goldin, P. R., McRae, K., Ramel, W., and Gross, J.J. "The Neural Bases of Emotion Regulation: Reappraisal and Suppression of Negative Emotion." *Biological Psychiatry* 63, no. 6 (March 15, 2008).

Gray, Rob. "Attending to the Execution of a Complex Sensorimotor Skill: Expertise Differences, Choking, and Slumps." *Journal of Experimental Psychology Applied* 10, no. 1 (March 2004): 42–54.

Grewen, K., Girdler, S. S., Amico, J., and Light, K. C. "Effects of Partner Support on Resting Oxytocin, Cortisol, Norepinephrine, and Blood Pressure Before and After Warm Partner Contact." *Psychosomatic Medicine* 67 (2005): 531–538.

Grinker, Roy R., and John P. Spiegel. *Men Under Stress.* Philadelphia: Blakiston, 1945.

Grosshandler, W. L., Bryner, N., Madrzykowski, D., and Kuntz, K. *Draft Report of the Technical Investigation of the Station Nightclub Fire.* National Institute of

Standards and Technology, March 2005. Available at http://www.nist.gov/
public_affairs/ncst/Front_matter_draft.pdf.

Grossman, Dave. *On Killing*. New York: Back Bay Books, 1995.

Gucciardi, D. F., and Dimmock, J. A. "Choking Under Pressure in Sensorimotor
Skills: Conscious Processing or Depleted Attentional Resources?" *Psychology
of Sport and Exercise* 9 (2008): 45–59.

Halberstam, David. *The Making of a Quagmire*. New York: Random House,
1965.

Hamilton, Linda, telephone interview with the author, February 26, 2009.

———. *The Person Behind the Mask*. Greenwich, CT: Abbex Publishing, 1997.

Hancock, Peter A., and James L. Szalma. *Performance Under Stress*. Aldershot,
UK: Ashgate Publishing, 2008.

Hanson, Victor Davis. *The Western Way of War*. Berkeley: University of California
Press, 1989.

Harmon-Jones, Eddie, and Piotr Winkielman, eds. *Social Neuroscience: Integrating
Biological and Psychological Explanations of Social Behavior*. New York: Guilford
Press, 2007.

Heimberg, Richard G. *Cognitive-behavioral Group Therapy for Social Phobia: Basic
Mechanisms and Clinical Strategies*. New York: Guilford Press, 2002.

Helmus, Todd C., and Russell W. Glenn. *Steeling the Mind: Combat Stress Reactions
and Their Implications for Urban Warfare*. Santa Monica, CA: Rand Corporation,
2005.

Hilton, Lisette. "Jansen Persevered Despite Olympic Disappointments." *ESPN.
com*. espn.go.com/classic/biography/s/Jansen_Dan.html/.

Hofmeyr, Ross. "Self-operation: Tracking down a good story."*doctorross.co.za*,
April 22, 2008. http://www.doctorross.co.za/antarctica/self-operation-
tracking-down-a-good-story.

IndieLondon. "Sleuth - Sir Michael Caine interview." *indielondon.co.uk*. http://
www.indielondon.co.uk/Film-Review/sleuth-sir-michael-caine-interview/.

Jacobs, Cindy, telephone interview with the author, January 23, 2009.

———. "Ceejays Survivor Story." *panicsurvivor.com*, Monday, 28 July 2008.
http://www.panicsurvivor.com/index.php/20080728661/Survivor-Stories/
Ceejays-Survivor-Story.html.

Jazzhaven Gallery. "Lion Monument." *jazzhaven.com*. http://www.gallery.
jazzhaven.com/lion-monument-6365i.htm.

Johnson, Rafer, ed. *Great Athletes*. Pasadena, CA: Salem Press, 2001.

Jones, Dan, and Alison Motluk. "Eight Ways To Get Exactly What You Want."
New Scientist, May 7, 2008.

Josephus, Flavius, trans. by William Whiston. *The Wars of the Jews*. Halifax, UK:
Milner and Sowerby, 1864.

Kahneman, Daniel, Paul Slovic, and Amos Tversky, eds. *Judgement under Uncertainty: Heuristics and Biases.* Cambridge, UK: Cambridge University Press, 1982.

Kavanagh, Jennifer. *Stress and Performance: A Review of the Literature and Its Applicability to the Military.* Santa Monica, CA: Rand Corporation, 2005.

Kirsch, P., et al. "Oxytocin Modulates Neural Circuitry for Social Cognition and Fear in Humans." *Journal of Neuroscience* 25 (2005): 11489–11493.

Kirsner, Kim, and Craig Speelman. *Implicit and Explicit Mental Processes.* Mahwah, NJ: Lawrence Erlbaum Associates, 1998.

Klein, Gary. *Sources of Power: How People Make Decisions.* Cambridge, MA: Massachusetts Institute of Technology Press, 1998.

Koch, John T. *Celtic Culture.* Santa Barbara, CA: ABC-CLIO, 2006.

Koltyn K. F., and W. P. Morgan. "Influence of Wet Suit Wear on Anxiety Responses to Underwater Exercise." *Undersea Hyperbaric Medicine* 24, no. 1 (1997): 23–28.

Kovaleski, Serge F., and Katie Zezima. "Students Recount Desperate Minutes Inside Norris Hall." *New York Times,* April 22, 2007.

La Rochefoucauld, Francois. *Maxims.* New York: Penguin Classics, 1982.

Leary, Mark R., and Robin M. Kowalski. *Social Anxiety.* New York: Guilford Press, 1997.

LeDoux, Joseph. *The Emotional Brain: The Mysterious Underpinnings of Emotional Life.* New York: Simon & Schuster, 1996.

Lieberman, H. R., Bathalon, G. P., Falco, C. M., Georgelis, J. H., Morgan, C. A., Niro, P., and Tharion, W. J. "The 'Fog of War': Documenting Cognitive Decrements Associated with the Stress of Combat." *Proceedings of the 23rd Army Science Conference,* December 2002.

Lieberman, Matthew, telephone interview with the author, March 12, 2009.

Lieberman, M. D., Eisenberger, E. I., Crockett, M. J., Tom, S. M., Pfeifer, J. H., and Way, B. M. "Putting Feelings Into Words: Affect Labeling Disrupts Amygdala Activity in Response to Affective Stimuli." *Psychological Science* 18, no. 5 (2007): 421–428.

Lieberman, Philip, and Stephen M. Kesslyn. *Human Language and Our Reptilian Brain.* Cambridge, MA: Harvard University Press, 2002.

Livingstone, David. *Missionary Travels and Researches in South Africa.* London: J. Murray, 1899.

Livy (Titus Livius), ed. by Ernest Rhys, trans. by Rev. Canon Roberts. *The History of Rome, Vol. I. Sections 2.12–13.* London: J. M. Dent, 1912.

Loach, D., Frischen, A., Bruce, N., and Tsotsos, J. K. "An Attentional Mechanism for Selecting Appropriate Actions Afforded by Graspable Objects." *Psychological Science* 19 no. 12 (2008): 1253–1257.

Longman, Jere. "900 Feet Up With Nowhere to Go but Down." *New York Times,* March 14, 2008.

Lubell, Sam. "On the Therapist's Couch, a Jolt of Virtual Reality." *New York Times,* February 19, 2004.

Lyubomirsky, Sonja, Lorie Sousa, and Rene Dickerhof. "The Costs and Benefits of Writing, Talking, and Thinking about Life's Triumphs and Defeats." *Journal of Personality and Social Psychology* 90, no. 4 (2006): 692–708.

Mangum, Geoff. "The Neurophysiology of Golf Putting." *Puttingzone.com.* http://www.puttingzone.com/Dystonia/yipsstudy.html/.

Marks, Isaac Meyer. *Fears, Phobias, and Rituals.* New York: Oxford University Press, 1987.

Marsh, Alton K. "'Attitude Flying' with Dick Rutan." *AOPA Pilot* (October 2008): 128–129.

Marshall, John R., and Suzanne Lipsett. *Social Phobia: From Shyness to Stage Fright.* New York: Basic Books, 1995.

McCabe, P. M., Schneiderman, N., Field, T. M., and Wellens, A. R., eds. *Stress, Coping, and Cardiovascular Disease.* Mahwah, NJ: Lawrence Erlbaum, 2000.

McCabe, Randi, telephone interview with the author, February 5, 2009.

McIntyre C. K., Miyashita, T., Setlow, B., Marjon, K. D., Steward, O., Guzowski, J. F., and McGaugh, J. L. "Memory-Influencing Intra-Basolateral Amygdala Drug Infusions Modulate Expression of Arc Protein in the Hippocampus." *Proceedings of the National Academy of Sciences* 102, no. 30 (2005): 10718–10723.

McNally, Ivan. "Contrasting Concepts of Competitive State-Anxiety in Sport: Multidimensional Anxiety and Catastrophe Theories." *Athletic Insight* 4, no. 2 (August, 2002): 10–22.

McNally, Richard J. *Panic Disorder: A Critical Analysis.* New York: Guilford Press, 1994.

McNaughton, Neil. *Biology and Emotion.* Cambridge, UK: Cambridge University Press, 1989.

Macphail, Euan M. *The Neuroscience of Animal Intelligence: From the Seahare to the Seahorse.* New York: Columbia University Press, 1993.

Mikulincer M., V. Florian, and G. Hirschberger. "The Existential Function of Close Relationships: Introducing Death Into the Science of Love." *Personality and Social Psychology Review* 7, no. 1 (2003): 20–40..

Mileti, D. S., Darlington, J. D. "The Role of Searching in Shaping Reactions to Earthquake Risk Information." *Social Problems* 44 (1997): 89–102.

Mobbs, D., Petrovic, P., Marchant, J. L., Hassabis, D., Weiskopf, N., Seymour, B., Dolan, R. J., and Frith, C. D. "When Fear Is Near: Threat Imminence Elicits Prefrontal-Periaqueductal Gray Shifts in Humans." *Science* 317 (August 24, 2007): 1079–1083.

Morgan, C. A., telephone interviews with the author, April 18, 2006 and February 1, 2007.

Morgan, C. A., J. H. Krystal, S. M. Southwick. "Toward Early Pharmacological Posttraumatic Stress Intervention." *Biological Psychiatry* 53, no. 9 (2003): 834–843.

Morgan, W. P. "Anxiety and Panic in Recreational Scuba Divers." *International Journal of Sports Medicine* 20, no. 6 (December 1995): 398–421.

Morgan, W. P., J. S. Raglin, and P. J. O'Connor. "Trait Anxiety Predicts Panic Behavior In Beginning Scuba Students." *International Journal of Sports Medicine* 25, no. 4 (May 2004): 314–322.

Morris, Tony, Peter Terry, and Sandy Gordon. *Sport and Exercise Psychology: An International Perspective.* Morgantown, WV: Fitness Information Technology, 2007.

Mujica-Parodi, L. R., Strey, H. H., Frederick, B., Savoy, R., Cox, D., Botanov, Y., Tolkunov, D., Rubin, D., and Weber, J. "Chemosensory Cues to Conspecific Emotional Stress Activate Amygdala in Humans." PLoS (in press).

Mujica-Parodi, L. R., Renelique, R., and Taylor, M. K. "Higher Body Fat Percentage Is Associated with Increased Cortisol Reactivity and Impaired Cognitive Resilience in Response to Acute Emotional Stress." *International Journal of Obesity* 33 (2009): 157–165.

Muraven, Mark, and Roy F. Baumeister. "Self-Regulation and Depletion of Limited Resources: Does Self-Control Resemble a Muscle?" *Psychological Bulletin* 126, no. 2 (2000): 247–259.

Murphy, Audie. *To Hell and Back.* New York: Henry Holt, 1949.

National Institute of Mental Health. "The Numbers Count: Mental Disorders in America." *nimh.nih.gov,* February 04, 2009. http://www.nimh.nih.gov/health/publications/the-numbers-count-mental-disorders-in-america/index.shtml.

National Transportation Safety Board. "Safety Recommendation A-96–15 through -20." Washington, D.C., May 31, 1996. *www.ntsb.gov/recs/letters/1996/a96_15_30.pdf.*

———. "Aircraft Accident Brief DCA07MA003." Washington, D.C., May 1, 2007.

———. "Aircraft Accident Brief NYC06FA205." Washington, D.C., September 27, 2007.

Neurophilosophy. "The baller's brain (and his pinky)." *scienceblogs.com/neurophilosophy/,* August 12, 2008. http://www.scienceblogs.com/neurophilosophy/2008/08/the_ballers_brain.php.

New Scientist. "Startle." *Newscientist.com,* September 2, 2006. *www.newscientist.com/article/mg19125672.300-the-word-startle.html/.*

Nicolaou, K. C., and Tamsyn Montagnon. *Molecules that Changed the World.* Weinheim, Germany: Wiley-VCH, 2008.

Olivier, Laurence. *Confessions of an Actor.* New York: Simon & Schuster, 1982.

Osgood, Charles Grosvenor, ed. *Boswell's Life of Johnson.* New York: Charles Scribner's Sons, 1917.

Otter, Johann, telephone interview with the author, June 1, 2007.

Pal, G. K., S. Velkumary, and Madanmohan. "Effect of Short-Term Practice of Breathing Exercises on Autonomic Functions in Normal Human Volunteers." *Indian Journal of Medical Research* 120 (August 2004): 115–121.

Panksepp, Jaak. *Affective Neuroscience.* New York: Oxford University Press, 1998.

Peurifoy, Reneau Z. *Anxiety, Phobias, and Panic: A Step-by-Step Program for Regaining Control of Your Life.* New York: Warner Books, 2005.

Pfaff, Donald W. *The Neuroscience of Fair Play.* New York: Dana Press, 2007.

Phillips, D. P., Liu, G. C., Kwok, K., Jarvinen, J. R., Zhang, W., and Abramson, I. S. "The Hound of the Baskervilles Effect: Natural Experiment on the Influence of Psychological Stress on Timing of Death." *BMJ* 323 (December 22–29, 2001): 1443–1446.

Plutarch, trans. by Aubrey Stewart and George Long. *Plutarch's Lives.* London: George Bell and Sons, 1900. Available at http://history-world.org/plut2.pdf.

Pool, Robert. "Why Do People Die That Way?" *New Scientist,* February 28, 2009.

Rachman, S. J. *Fear and Courage,* 2d ed. New York: W. H. Freeman, 1990.

Rathelot, Jean-Alban, and Peter L. Strick. "Subdivisions of Primary Motor Cortex Based on Cortico-Motoneuronal Cells." *Proceedings of the National Academy of Sciences* 106, no. 3 (2009): 918–23.

Rejan, Wendy. "Shooting Without Bullets." *Monmouth Message,* August 8, 2008.

Ripley, Amanda. *The Unthinkable: Who Survives When Disaster Strikes—and Why.* New York: Crown Publishers, 2008.

Rogozov, L. I. "Self Operation." *Soviet Antarctic Expedition Information Bulletin* 4 (1964): 223–224.

Rohrmann, Sonja. "Changing Psychobiological Stress Reactions by Manipulating Cognitive Processes." *International Journal of Psychophysiology* 33, no. 2 (August 1, 1999): 149–161.

Rostker, B. D., Hanser, L. M., Hix, W. M., Jensen, C., Morral, A. R., Ridgeway, G., and Schell, T. L. *Evaluation of the New York City Police Department Firearm Training and Firearm-Discharge Review Process.* Santa Monica, CA: Rand Corporation, 2008.

Russian Antarctic Expedition. "Russian Antarctic Stations—Overview." *aari.aq.* http://www.aari.aq/default_en.html.

Sato, W., Yoshikawa, S., Kochiyama, T. and Matsumura, M. "The Amygdala Processes the Emotional Significance of Facial Expressions: An fMRI Investigation Using the Interaction between Expression and Face Direction." *Neuroimage* 22, no. 2 (June 2004): 1006–1013.

Scott, Matt. "Olympics: Korean Double Medallist Expelled for Drug Use." *Guardian* (UK), August 15, 2008.

Shakespeare, William. *Henry V.* Available at http://www.william-shakespeare. info/shakespeare-play-king-henry-v.htm.

Shibusawa, Naoko. *America's Geisha Ally: Reimagining the Japanese Enemy.* Cambridge, MA: Harvard University Press, 2006.

Siddle, Bruce K. *Sharpening the Warrior's Edge.* Belleville: PPCT Research Publications, 2003.

Smith, Michael David. "Andy Bolton Deadlifts 1,008 Pounds." *Backporch.fanhouse. com,* April 7, 2009. http://www.backporch.fanhouse.com/2009/04/07/andy-bolton-deadlifts-1-008-pounds/.

Spectre, Peter H. *A Mariner's Miscellany.* London: Seafarer Books, 2005.

Spencer, Richard. "Australian Bush Fires' Street Of Death: How Fire Consumed Pine Ridge Road, Kinglake." *Telegraph* (UK), February 14, 2009.

Staal, Mark A. *Stress, Cognition, and Human Performance: A Literature Review and Conceptual Framework.* Moffett Field, CA: National Aeronautics and Space Administration Ames Research Center, 2004.

Stadtler, Mike. *The Psychology of Baseball: Inside the Mental Game of the Major League Player.* New York: Gotham Books, 2007.

Starkey, Joe. "Blass Earns Win over 'Disease.'" *Pittsburgh Tribune-Review,* July 6, 2008.

Stein, Murray B., and Dan J. Stein. "Social Anxiety Disorder." *Lancet* 371, no. 9618 (March 29, 2008): 1115–1125.

Stetson C., M. P. Fiesta, and D.M. Eagleman. "Does Time Really Slow Down During a Frightening Event?" *PLoS ONE* 2, no. 12 (2007). http://www.plosone.org/article/info:doi/10.1371/journal.pone.0001295/.

Szalavitz, Maia. "'Cuddle Chemical' Could Treat Mental Illness." *New Scientist,* May 14, 2008.

Taylor, Marc, telephone interview with the author, January 31, 2007.

Taylor, S. E., Burklund, L. J., Eisenberger, N. I., Lehman, B. J., Hilmert, C. J., and Lieberman, M. D. "Neural Bases of Moderation of Cortisol Stress Responses by Psychosocial Resources." *Journal of Personality and Social Psychology* 95, no. 1 (2008): 197–211.

Thomas, Ian, telephone interview with the author, March 12, 2009.

Thomas, Katie. "The Secret Curse of Expert Archers." *New York Times,* August 1, 2008.

Thompson, L. A., Williams, K. L., and L'Esperance, P. R. "Context-Dependent Memory under Stressful Conditions: The Case of Skydiving." *Human Factors,* December 22, 2001.

Tindall, Blair. "Better Playing Through Chemistry." *New York Times,* October 17, 2004.

Troiani, D., Krick, R. K., Knoke, K., and White, L. *Don Troiani's American Battles: The Art of the Nation at War, 1754–1865.* Mechanicsburg, PA: Stackpole Books, 2006.

Tsien, Joe Z. "The Memory Code." *Scientific American,* July 2007, 52–59.

Tully, Keith, Yan Li, and Vadim Y. Bolshakov. "Keeping in Check Painful Synapses in Central Amygdala." *Neuron* 56, no. 5 (2007): 757–759.

United States Census Bureau. "Census Bureau Estimates Number of Adults, Older People and School-Age Children in States." *census.gov,* March 10, 2004. http://www.census.gov/Press-Release/www/releases/archives/population/001703.html.

University of California. "Neuroscientists identify how trauma triggers long-lasting memories." *universityofcalifornia.edu,* July 26, 2005. http://www.universityofcalifornia.edu/news/article/7355.

Vetter, Craig. "Bill Stone in the Abyss." *Outside,* November 1994.

Vegetius Renatus, Flavius, trans. by John Clarke. *The Military Institutions of the Romans [De Re Militari].* The Military Service Publishing Company, 1944. Available at http://www.sacred-archery.com/De_Re_Militari.pdf/.

Viegas, Domingos Xavier. "Anatomy of a Blow-Up." Wildfire, September-October 2006.

Violand, Clay. "Virginia Tech Shootings: A Survivor's Account."*WTOP.com,* April 18, 2007. http://www.wtop.com/?sid=1118941&nid=25/.

Walkup, J. T., Mink, J. W., and Hollenbeck, P. J., eds. *Tourette Syndrome.* Philadelphia: Lippincott Williams & Wilkins, 2006.

Wegner, Daniel. *The Illusion of Conscious Will.* Cambridge, MA: Massachusetts Institute of Technology Press, 2002.

Wells, Chris, telephone interview with the author, February 18, 2009.

Williams, Caroline. "The 25 Hour Day." *New Scientist,* February 4, 2006, 34–37.

Williams, Neil. *Aerobatics.* Shrewsbury, England: Airlife Publishing, 1975.

———. *Airborne.* Shrewsbury, England: Airlife Publishing, 1977.

———. "Structural failure." *FLIGHT International* (June 19, 1970): 993–994.

Wilson, Michael. "50 Shots Fired, and the Experts Offer a Theory." *New York Times,* November 27, 2006.

Wilson, Timothy D. *Strangers to Ourselves: Discovering the Adaptive Unconscious.* Cambridge, MA: Belknap Press, 2002.

Wise, Jeff. "Taking the Scream Test." *Popular Mechanics,* August 2008, 52–54.

————. "I'll Try Anything." *Popular* Mechanics, August 2009, 44–46.

————. "The Unluckiest Ship in the Navy." *Popular Mechanics,* November 2008, 90–93.

Wolff, Alexander. "Whooosh!" *Sports Illustrated,* February 28, 1994, 18–23.

Woodman, Tim, telephone interview with the author, February 7, 2007.

Woodman, T. and Hardy, L. "Dynamic Systems, Catastrophe Models, and Performance." *Science et Motricite,* no. 60 (2007): 63–68.

Yellowtail, Susan, telephone interview with the author, March 19, 2007.

Yerkes, Robert M., and John D. Dodson. "The Relation of Strength of Stimulus to the Rapidity of Habit-Formation." *Journal of Comparative Neurology and Psychology* 18 (1908): 459–482.

Zatsiorsky, Vladimir M., and William J. Kraemer. *Science and Practice of Strength Training,* 2d ed. Champaign, IL: Human Kinetics, 2006.

INDEX